Multiscale Modeling in Nanophotonics

Multiscale Modeling in Nanophotonics

Materials and Simulation

Alexander Bagaturyants
Mikhail Vener

Pan Stanford Publishing

Published by

Pan Stanford Publishing Pte. Ltd.
Penthouse Level, Suntec Tower 3
8 Temasek Boulevard
Singapore 038988

Email: editorial@panstanford.com
Web: www.panstanford.com

British Library Cataloguing-in-Publication Data
A catalogue record for this book is available from the British Library.

Multiscale Modeling in Nanophotonics:
Materials and Simulation

ISBN 978-981-4774-40-6 (Hardcover)
ISBN 978-1-315-10976-3 (eBook)

The more accurate the calculations become, the more the concepts tend to vanish into thin air.

—R. S. Mulliken (Nobel Prize in Chemistry, 1966),
J. Chem. Phys., **43**, S2 (1965)

Contents

Preface

The idea of predicting the useful properties of various materials on the basis of multiscale simulations has become quite popular in recent years. Of special interest are nanostructured materials, which have a hierarchical structure. Among them, organic functional materials are promising and are considered materials of the future because of their flexibility and versatility. Their functional properties are inherited from the functional molecule that is at the heart of the hierarchical structure. On the other hand, the properties of the functional molecule, in particular its absorption and emission spectra, strongly depend on its interactions with its molecular environment. Therefore, multiscale simulation methods used for predicting the properties of organic functional materials should be atomistic, that is, they should be based on (classical and/or quantum) methods that explicitly take into account the molecular structure and intermolecular interactions on an atomic level.

This book is devoted to an exposition of these atomistic methods and their use in practical simulations. It contains six chapters: (1) Introduction, (2) Basics of Classical Mechanics, (3) Quantum Chemistry Methods for Molecular Disordered Materials, (4) Classical Methods of Molecular Simulations of Disordered Materials, (5) Vibronic Interactions, and (6) Atomistic Simulation of Hierarchical Nanostructured Materials for Optical Chemical Sensing. A. Bagaturyants is responsible for Chapters 1, 2, 3, 5, and for the general structure and composition of the book, and Chapter 4 is written by M. Vener. Dr. A. Scherbinin carefully read Chapter 5 and made many valuable comments and amendments.

A. Bagaturyants acknowledges financial support from the Russian Science Foundation project no. 14-43-00052 for the writing and preparation of this book.

Alexander Bagaturyants and Mikhail Vener

Moscow

February 5, 2017

Foreword

Nanophotonics considers photonic phenomena at the nanoscale and is used in a wide variety of applications ranging from biochemistry

to electrical engineering. These applications require specific so-called functional materials specially designed for a particular purpose. Among them, the most important and versatile are organic functional materials. To select the best composition and structure of a given material, it is important to theoretically predict its photonic properties, because blind experiment can be successful only occasionally. This goal is achieved by using atomistic material simulations, which predict the properties of a material on the basis of its atomistic composition and structure. This book, written by well-known specialists in theoretical chemistry, presents methods that are currently used in modeling nanophotonic materials and devices. It encompasses classical, quantum, and statistical methods and presents examples of their practical applications in optical chemical sensing. It might be recommended to advanced and graduate-level students, researchers and technologists involved in materials science, and all other readers interested in atomistic simulations in different fields.

Prof. Michael Alfimov

Full member of the Russian Academy of Sciences

Chapter 1

Introduction

Interest in functional organic materials for applications in photonics goes back to the eighties of the previous century, when organic light-emitting devices (OLEDs) [1] and organic solar cells (photovoltaic devices) were first designed [2]. The main advantages of organic materials are their great variability, versatility, low cost, and ease of production. The main component of such functional materials is a functional molecule, which determines their functionality and their useful optical and electronic properties.

A material might be fully composed of one functional compound, or it might be designed as a matrix doped with some sensitizing agent. This latter design is typical for light-emitting phosphorescent materials [3, 4], photovoltaic materials [5, 6], and optical sensing materials [7]. Functional organic materials for applications in photonics are typical disordered semiconductors, which are commonly used in devices as layers (about 10–40 nm thick). The physical properties of functional molecules in such layers depend on their local environment, which, in turn, depends on the morphology and microstructure of the material. The structure of a functional organic material exhibits several levels of organization: atomic-molecular (1–2 nm), supramolecular (2–10 nm), nanosized (10–100 nm), and microscopic (100–1000 nm and more).

A theoretical understanding and prediction of properties of the material, which is necessary for optimal selection of its constituents,

Multiscale Modeling in Nanophotonics: Materials and Simulation
Alexander Bagaturyants and Mikhail Vener

ISBN 978-981-4774-40-6 (Hardcover), 978-1-315-10976-3 (eBook)
www.panstanford.com

should be based on modeling that directly takes into account the atomistic structure of the material and its hierarchical multiscale organization. The first works on atomistic multiscale simulation (AMS) of materials appeared only in the late 1990s, while the first work in which the AMS strategy was applied to organic materials was published only in 2001 [8]. It seems that the term "atomistic multiscale simulation" was coined in Ref. [9].

This book is devoted to AMS of disordered (amorphous) organic materials for photonics applications. AMS here involves two levels: (i) modeling of the microstructure of the disordered (amorphous) material (layer) using molecular dynamics (MD) methods with classical atomistic potentials and (ii) modeling the electronic properties of a functional molecule embedded in a supramolecular cluster (including its neighboring molecules) using molecular quantum-mechanical methods. The book contains three main parts: (I) Quantum Chemistry Methods for Disordered (Organic) Materials, (II) Classical Molecular Dynamics Methods for Disordered (Amorphous) Materials, and (III) Multiscale Strategy and Applications. The main content is preceded by a concise exposition of basics of classical mechanics in a separate chapter. Classical mechanics lies in the ground of both quantum-chemical and MD methods, and what is even more important, many molecular properties (structural, vibrational, and others) are described in the framework of purely classical concepts.

The book is an intermediate between a monograph and a textbook for advanced PhD students.

References

1. C. W. Tang and S. A. VanSlyke, Organic electroluminescent diodes, *Appl. Phys. Lett.*, 1987, **51**, 913–915.
2. J. Hou and X. Guo, Active layer materials for organic solar cells, in *Organic Solar Cells: Materials and Device Physics*, W. C. H. Choy (ed.), Springer, London, 2013, pp. 17–42; C. W. Tang, Two-layer organic photovoltaic cell, *Appl. Phys. Lett.*, 1986, **48**, 183–185; G. A. Chamberlain, Organic solar cells: a review, *Sol. Cells*, 1983, **8**, 47–83.
3. S. Kappaun, C. Slugovc, and E. J. W. List, Phosphorescent organic light-emitting devices: working principle and iridium based emitter materials, *Int. J. Mol. Sci.*, 2008, **9**, 1527–1547.

4. M. A. Baldo, M. E. Thompson, and S. R. Forrest, High-efficiency fluorescent organic light-emitting devices using a phosphorescent sensitizer, *Nature*, 2000, **403**, 750–753.

5. N. E. Jackson, B. M. Savoie, T. J. Marks, L. X. Chen, and M. A. Ratner, The next breakthrough for organic photovoltaics?, *J. Phys. Chem. Lett.*, 2015, **6**, 77–84.

6. H. Phillips, Z. Zheng, E. Geva, and B. D. Dunietz, Orbital gap predictions for rational design of organic photovoltaic materials, *Org. Electron.*, 2014, **15**, 1509–1520.

7. A. Bagaturyants, M. Alfimov, Atomistic simulation of hierarchical nanostructured materials for optical chemical sensing, in *Chemical Sensors: Simulation and Modeling*, Vol. 4, *Optical Sensors*, G. Korotcenkov (ed.), Momentum Press, New York, 2013, pp. 1–38.

8. W. A. Goddard III, T. Cagin, M. Blanco, N. Vaidehi, S. Dasgupta, W. Floriano, M. Belmares, J. Kua, G. Zamanakos, S. Kashihara, M. Iotov, and G. Gao, Strategies for multiscale modeling and simulation of organic materials: polymers and biopolymers, *Comput. Theor. Polym. Sci.*, 2001, **11**, 329–343.

9. S. Emelyanova, V. Chashchikhin, and A. Bagaturyants, Modeling of the structure and properties of an amorphous OXD-7 layer, *Chem. Phys. Lett.*, 2013, **590**, 101–105.

Chapter 2

Basics of Classical Mechanics

We start from *classical mechanics* for many reasons. The first and evident reason is that it forms the basis of molecular dynamics simulations. The second reason is that an understanding of *quantum chemistry* is impossible without addressing classical mechanics. This reason is deeply rooted in the essence of quantum chemistry as a science. Quantum chemistry can be considered as an application of quantum theory to chemistry. In fact, quantum chemistry was developed by physicists (W. Heitler and D. London [1]) in their daring attempt to explain the nature of chemical bonds in the simplest chemical molecule H_2. Subsequently, quantum chemistry in many respects developed as an application of quantum mechanics to chemical and, par excellence, molecular problems. By now, computational molecular quantum mechanics (including methods and applications) still remains one of the main areas of quantum-chemical activity. Hence, quantum mechanics is one of the main foundations of quantum chemistry, and studying quantum chemistry is impossible without a knowledge of the basics of quantum mechanics. However, according to the correspondence principle, the basics of quantum mechanics and quantum-mechanical equations cannot be properly formulated and fully understood without classical mechanics. Hence, the second reason is now evident.

The third reason is even more important and principal. From the very start, it became evident that chemical systems and processes are

Multiscale Modeling in Nanophotonics: Materials and Simulation
Alexander Bagaturyants and Mikhail Vener

ISBN 978-981-4774-40-6 (Hardcover), 978-1-315-10976-3 (eBook)
www.panstanford.com

so complicated that their understanding cannot be simply reduced to solving exercises in molecular quantum mechanics. Moreover, such a fundamental chemical notion as the *chemical structure* is purely classical. The existence of different stable chemical isomers or different isomeric chemical systems, mechanisms of chemical reactions and reaction paths, molecular vibrations and spectroscopy, and many other chemical phenomena and concepts cannot be treated without passing to a purely classical level of description, that is, approaches based on classical mechanics. Therefore, quantum chemistry is inevitably based on a mixture of quantum and classical concepts, and because of this, modern theoretical methods of studying chemical problems at a molecular level involve both quantum and classical approaches.

2.1 Particles, Trajectories, Generalized Coordinates, and Newtonian and Lagrangian Equations

2.1.1 Introduction: Trajectories

One of the main idealizations used in physics is the concept of a material point within which the real size of a physical body is neglected. However, we know that the world is constructed of particles. Therefore, when considering a physical body at a microscopic level, one may use this idealization quite reasonably.

In chemistry, the main particles are electrons and nuclei, which compose atoms and molecules. Molecules play the most important role in chemistry, being its main actors. The lightest nucleus, the proton, is approximately 2000 times heavier than an electron. The entire mass of an atom is actually almost fully concentrated in its nucleus, so a molecule or a molecular system, from the mechanical point of view, might be quite reasonably considered as a system composed of material points (atoms, whose position can formally be associated with the positions of their nuclei).

Furthermore, in classical mechanics, it is assumed that at any instant of time an observer can locate a particle at a mathematical point of the whole space specified by its radius vector $\mathbf{r}(t)$ with Cartesian coordinates $x(t), y(t), z(t)$.

The function $\mathbf{r}(t)$ is the trajectory of the particle. The time derivative of $\mathbf{r}(t)$ $\mathbf{v} = \dot{\mathbf{r}} = \frac{d\mathbf{r}(t)}{dt}$ with components $\left(\frac{dx}{dt}, \frac{dy}{dt}, \frac{dz}{dx}t\right) = (\dot{x}, \dot{y}, \dot{z})$ is the vector velocity of the particle, and the second derivative $\frac{d^2\mathbf{r}}{dt^2} = \ddot{\mathbf{r}}$ is its vector acceleration. Here, a dot over the variable designates, as usual, total differentiation with respect to time. The trajectory $\mathbf{r}(t)$ fully describes the time evolution of the particle and, hence, its mechanical state.

2.1.2 Newtonian Equations

The motion of a particle along its trajectory is determined by the Newtonian equations of motion

$$m\ddot{x} = f_x,\ m\ddot{y} = f_y,\ m\ddot{z} = f_z, \tag{2.1a}$$

or in vector form

$$m\frac{d^2\mathbf{r}}{dt^2} = \mathbf{f}, \tag{2.1b}$$

where m is the particle mass and $\mathbf{f}$ is the vector force acting on the particle with three Cartesian components f_x, f_y, f_z.

If a system consists of a large (but finite) number of particles N, the time evolution of their positions and, hence, the mechanical state of the system are described by N radius vectors $\mathbf{r}_k(t)$. The full set of radius vectors $\mathbf{r}_k(t)$ can be treated as the trajectory of the mechanical system of particles in the multi($3N$)-dimensional space of coordinates of all particles (so-called configuration space). The corresponding Newtonian equations for a system of particles can be written as follows:

$$m_k\frac{d^2\mathbf{r}_k}{dt^2} = \mathbf{f}_k \ (1 \le k \le N) \tag{2.1c}$$

Here, m_k is the mass of the kth particle, and the system is described by $3N$ Cartesian coordinates; that is, it possesses $3N$ degrees of freedom.

2.1.3 Generalized Coordinates

Cartesian coordinates are not always the best choice. In many cases, it is convenient to use a different coordinate system. In fact, any $3N$ parameters that uniquely define the position of the system can be used as a coordinate system. Such parameters q_i ($1 \le i \le 3N$) are called generalized coordinates. Note that generalized coordinates are not generally speaking related to individual particles anymore, and Cartesian coordinates are only a particular case of generalized coordinates. For uniformity and simplicity, it is convenient to use a consecutive numbering (r_i, $1 \le i \le 3N$) for Cartesian coordinates in the same way as for generalized coordinates wherever this is not confusing. Then, the Newtonian equations in Cartesian coordinates will be written as

$$m_i \frac{d^2 r_i}{dt^2} = f_i \ (1 \le i \le 3N), \tag{2.2}$$

where now m_i assumes the same value for a triad of numbers $i = 3k - 2, 3k - 1, 3k$ (formally, m_i is the mass of the kth particle, with k being the integer of $\frac{i+2}{3}$). Note here that the Newtonian equations are ordinary second-order differential equations. Therefore, their solutions are determined by $6N$ initial conditions, that is, by $3N$ coordinates r_{0i} and $3N$ velocities $\dot{r}_{0i}$ specified at a certain instant of time t_0. This means that in classical mechanics the entire history of a system of particles (i.e., its mechanical state) is completely determined by the $3N$-dimensional vectors of positions and velocities of its particles. A particular selection of all the coordinates of the system (generalized or Cartesian) will be called below its configuration.

Generalized coordinates can be expressed as a (generally speaking, nonlinear) function of Cartesian coordinates (and vice versa), so $q_i = q_i(x_k, y_k, z_k) = q_i(r_j)$, ($1 \le i, j \le 3N$) and ($1 \le k \le N$). Note that the generalized velocities $\dot{q}_i$ are linear functions of velocities $\dot{r}_i$ by virtue of the known rule of chain differentiation:

$$\dot{q}_i = \frac{dq_i}{dt} = \sum_j \frac{\partial q_i}{\partial r_j} \frac{dr_j}{dt} = \sum_j \frac{\partial q_i}{\partial r_j} \dot{r}_j, \tag{2.3}$$

and similarly

$$\dot{r}_i = \frac{dr_i}{dt} = \sum_j \frac{\partial r_i}{\partial q_j}\frac{dq_j}{dt} = \sum_j \frac{\partial r_i}{\partial q_j}\dot{q}_j. \tag{2.4}$$

From Eqs. 2.3 and 2.4 some other useful relations can be derived. First, we calculate the partial derivative of $\dot{r}_i$ with respect to $\dot{q}_i$. It follows directly from Eq. 2.4 that

$$\frac{\partial \dot{r}_i}{\partial \dot{q}_j} = \frac{\partial r_i}{\partial q_j}. \tag{2.5}$$

Next, we calculate the total derivative of $\frac{\partial r_i}{\partial q_j}$ with respect to time, first using the chain rule of differentiation (Eq. 2.3), then changing the order of differentiation, and finally again using Eq. 2.3:

$$\frac{d}{dt}\frac{\partial r_i}{\partial q_j} = \sum_k \frac{\partial}{\partial q_k}\left(\frac{\partial r_i}{\partial q_j}\right)\dot{q}_k = \sum_k \frac{\partial}{\partial q_j}\left(\frac{\partial r_i}{\partial q_k}\right)\dot{q}_k$$

$$= \frac{\partial}{\partial q_j}\sum_k \left(\frac{\partial r_i}{\partial q_k}\right)\dot{q}_k = \frac{\partial \dot{r}_i}{\partial q_j}, \tag{2.6a}$$

that is,

$$\frac{d}{dt}\frac{\partial r_i}{\partial q_j} = \frac{\partial \dot{r}_i}{\partial q_j}. \tag{2.6b}$$

Note here that a transformation of coordinates, for example, from Cartesian to generalized, or back) is characterized by the Jacobian of this transformation, namely

$$\mathbf{J}(r \to q) = \left[\frac{\partial q_i}{\partial r_j}\right],\ \mathbf{J}(q \to r) = \left[\frac{\partial r_i}{\partial q_j}\right],$$

$$\mathbf{J}(r \to q) = \mathbf{J}(q \to r)^{-1} \tag{2.7}$$

because $\sum_j \frac{\partial q_i}{\partial r_j}\frac{\partial r_j}{\partial q_k} = \delta_{ik}$.

2.1.4 Equation of Motion in Generalized Coordinates and Lagrangian Formalism

Now, we are in a position to derive the equation of motion in generalized coordinates. This is most conveniently done within

Lagrangian formalism. We first introduce it for the special case of Cartesian coordinates. Let us define the kinetic energy of the system as

$$T(\dot{r}_i)=\frac{1}{2}\sum_i m_i\dot{r}_i^2 \ . \tag{2.8}$$

It is evident that the Newtonian equations (Eq. 2.2) can be rewritten in the form

$$\frac{d}{dt}\frac{\partial T}{\partial \dot{r}_i}=f_i\,,(1\leq i\leq 3N)\,. \tag{2.9}$$

Now, we introduce an important assumption that our system of particles is conservative;[a] that is, the forces f_i acting on each particle can be obtained from a potential energy function $V(r_i)$ as follows:

$$f_i=-\frac{\partial V}{\partial r_i}\,,(1\leq i\leq 3N) \tag{2.10}$$

Note here that while the kinetic energy depends only on the velocities $\dot{r}_i$, the potential energy function depends only on the coordinates r_i. With this assumption, we further obtain

$$\frac{d}{dt}\frac{\partial T}{\partial \dot{r}_i}+\frac{\partial V}{\partial r_i}=0\ \ (1\leq i\leq 3N). \tag{2.11}$$

We define the Lagrangian function as

$$L(r_i,\dot{r}_i)=T(\dot{r}_i)-V(r_i)\,, \tag{2.12}$$

so we finally obtain the equations of motion for the system of particles in Lagrangian form

$$\frac{d}{dt}\frac{\partial L}{\partial \dot{r}_i}-\frac{\partial L}{\partial r_i}=0\,,(1\leq i\leq 3N). \tag{2.13}$$

These equations are equivalent to the Newtonian equations (Eq. 2.2). However, as distinct from those, the Lagrangian equations are invariant with respect to any transformation of coordinates and retain their form in any (generalized) coordinate system. In addition, the Lagrangian function $L(r_i,\dot{r}_i)$ possesses many useful properties.

First, we introduce generalized forces. Consider a small (infinitesimal) arbitrary displacement of the system $r_i \to r_i + \delta r_i$. This displacement in generalized coordinates can be expressed as

[a]A conservative system is a system in which the work done by a force is (i) independent of path, (ii) equal to the difference between the final and initial values of the energy function, and (iii) completely reversible.

$$\delta r_i = \sum_j \frac{\partial r_i}{\partial q_j} \delta q_j . \tag{2.14}$$

If we consider the work done on the system of particles by the forces f_i upon this displacement, we will obtain with the use of Eq. 2.14

$$\delta W = \sum_i f_i \delta r_i = \sum_i \sum_j f_i \frac{\partial r_i}{\partial q_j} \delta q_j = \sum_j F_j \delta q_j . \tag{2.15}$$

Equation 2.15 allows us to define the generalized force

$$F_j = \sum_i f_i \frac{\partial r_i}{\partial q_j} \tag{2.16}$$

acting along the generalized coordinate q_j. This definition is quite consistent with the concept of force in classical mechanics, because the generalized force expresses the work done on the system in the course of a unit displacement along a generalized coordinate. We may also quite easily represent the kinetic energy $T(\dot{r}_i)$ in generalized coordinates using Eq. 2.4

$$\begin{aligned} T(\dot{r}_i) &= \frac{1}{2}\sum_i m_i \dot{r}_i^2 = \frac{1}{2}\sum_i m_i \left(\sum_j \frac{\partial r_i}{\partial q_j} \dot{q}_j \right)^2 \\ &= \frac{1}{2}\sum_i m_i \left(\sum_j \frac{\partial r_i}{\partial q_j} \dot{q}_j \right) \left(\sum_j \frac{\partial r_i}{\partial q_k} \dot{q}_k \right) \\ &= \frac{1}{2}\sum_{j,k} \left(\sum_i m_i \frac{\partial r_i}{\partial q_j} \frac{\partial r_i}{\partial q_k} \right) \dot{q}_j \dot{q}_k , \end{aligned}$$

and we obtain

$$T(q,\dot{q}) = \frac{1}{2}\sum_{ij} m_{ij}(q) \dot{q}_i \dot{q}_j , \tag{2.17}$$

where we used the designation

$$m_{jk}(q) = \sum_i m_i \frac{\partial r_i}{\partial q_j} \frac{\partial r_i}{\partial q_k} . \tag{2.18}$$

It is evident that

$$m_{jk}(q) = m_{kj}(q). \tag{2.19}$$

The kinetic energy in generalized coordinates is no longer a function of only velocities but also depends on coordinates.

Now, we start with Eq. 2.3 and transform it from Cartesian to generalized coordinates by multiplying both parts of this equation by $\frac{\partial r_i}{\partial q_j}$ and summing over i. For the right-hand side this immediately gives

$$\sum_i f_i \frac{\partial r_i}{\partial q_j} = F_j \,. \tag{2.20}$$

The left-hand side of Eq. 2.20 can be transformed as follows:

$$\sum_i \frac{\partial r_i}{\partial q_j}\frac{d}{dt}\frac{\partial T}{\partial \dot{r}_i} = \sum_i \frac{d}{dt}\left(\frac{\partial T}{\partial \dot{r}_i}\frac{\partial r_i}{\partial q_j}\right) - \sum_i \frac{\partial T}{\partial \dot{r}_i}\frac{d}{dt}\left(\frac{\partial r_i}{\partial q_j}\right) \tag{2.21}$$

Then, we use Eq. 2.4 to modify the first term in the right-hand side of Eq. 2.21 and Eq. 2.5 to modify the second term. This gives

$$\sum_i \frac{d}{dt}\left(\frac{\partial T}{\partial \dot{r}_i}\frac{\partial r_i}{\partial q_j}\right) = \sum_i \frac{d}{dt}\left(\frac{\partial T}{\partial \dot{r}_i}\frac{\partial \dot{r}_i}{\partial \dot{q}_j}\right) = \frac{d}{dt}\frac{\partial T}{\partial \dot{q}_j} \tag{2.22}$$

and

$$\sum_i \frac{\partial T}{\partial \dot{r}_i}\frac{d}{dt}\left(\frac{\partial r_i}{\partial q_j}\right) = \sum_i \frac{\partial T}{\partial \dot{r}_i}\frac{\partial \dot{r}_i}{\partial q_j} = \frac{\partial T}{\partial q_j} \,. \tag{2.23}$$

Finally, we obtain

$$\frac{d}{dt}\frac{\partial T}{\partial \dot{q}_j} - \frac{\partial T}{\partial q_j} = F_j \,. \tag{2.24}$$

Consider again only a conservative system. In this case, the generalized forces F_j are given by the same equation as in the case of Cartesian coordinates (Eq. 2.10), because by definition (Eq. 2.20)

$$F_j = \sum_i f_i \frac{\partial r_i}{\partial q_j} = \sum_i -\frac{\partial V}{\partial r_i}\frac{\partial r_i}{\partial q_j} = -\frac{\partial V}{\partial q_j}, \ (1 \le j \le 3N) \,. \tag{2.25}$$

And again, we introduce the Lagrangian function as $L(q,\dot{q}) = T(q,\dot{q}) - V(q)$. Note that because V does not depend on $\dot{q}$, $\frac{\partial V}{\partial \dot{q}} = 0$. Therefore, combining Eqs. 2.24 and 2.25, we finally obtain the desired equations of motion in generalized coordinates in the

Lagrangian formulation as follows:

$$\frac{d}{dt}\frac{\partial L}{\partial \dot{q}_i} - \frac{\partial L}{\partial q_i} = 0, \ (1 \le i \le 3N) \qquad (2.26)$$

We see that these equations have the same form as the Lagrangian equations of motion in Cartesian coordinates, which proves our statement that Lagrangian equations of motion are invariant to coordinate transformations.

2.1.5 Momentum

We introduce here a new dynamical variable called the momentum and define its components:

$$p_i = \frac{\partial L}{\partial \dot{q}_i}, \ (1 \le i \le 3N) \qquad (2.27)$$

The momentum components p_i are related to each generalized coordinate, and q_i and p_i are called conjugate variables. In terms of these quantities, the Lagrangian equations (Eqs. 2.12 and 2.13) take the form

$$\frac{d}{dt} p_i \equiv \dot{p}_i = F_i = \frac{\partial L}{\partial q_i}, \ (1 \le i \le 3N). \qquad (2.28)$$

Because the potential energy function is independent of velocities, momentum can be presented in the form

$$p_i = \frac{\partial T}{\partial \dot{q}_i} = \sum_j m_{ij} \dot{q}_j, \ (1 \le i \le 3N), \qquad (2.29)$$

where the coefficients $m_{ij}(q) = m_{ji}(q)$ are determined by Eq. 2.18.

In Cartesian coordinates, we may introduce the momentum vector of an individual particle $\mathbf{P}_k$ with components p_{kx}, p_{ky}, p_{kz}. From the definition of the Lagrangian function in Cartesian coordinates (Eq. 2.7), it readily follows that

$$\mathbf{p}_k = m_k \mathbf{v}_k. \qquad (2.30)$$

2.2 The Principle of Least Action

2.2.1 Action

We introduce now a quantity called the action integral or, simply, action:

$$S(t_1,t_2)=\int_{t_1}^{t_2} L(q,\dot{q},t)\,dt\,, \tag{2.31}$$

where we mean under q the entire set of generalized coordinates $\{q_i\}$ of a system of particles under consideration. The value of S depends on the trajectory $q(t)$.

2.2.2 The Principle of Least Action: Derivation

Let the coordinate of the system at the instants of time t_1 and t_2 assume the values $q(t_1) = q_1$ and $q(t_2) = q_2$. The principle of least action (better to say, stationary action) states that the motion of the system between these two positions is carried out in such a way that the action assumes a minimum (stationary) value for the trajectory of the system $q(t)$. Let $q^0(t)$ be a trajectory that satisfies the equations of motion (Eq. 2.23) and passes through two points q_1 and q_2 at the instants of time t_1 and t_2. Consider now a trajectory $q(t)$ that differs from $q^0(t)$ by a small (infinitesimal) variation $\delta q(t)$, $q(t) = q^0(t) + \delta q(t)$. The admissible variations $\delta q(t)$ must obey the condition that $\delta q(t_1) = \delta q(t_2) = 0$, because all the admissible trajectories $q(t)$ must assume the same fixed values q_1 and q_2 at t_1 and t_2. According to the principle of least action, any admissible variation of $q^0(t)$ should lead to a zero variation of $S(t_1,t_2)$; that is, the principle of least action in the mathematical form states that

$$\delta S(t_1,t_2)=\delta\int_{t_1}^{t_2} L(q,\dot{q},t)dt=0 \tag{2.32}$$

for any admissible variation of the trajectory $\delta q(t)$.

Consider now the variation of the action integral $S(t_1,t_2)$ due to the variation $\delta q(t)$of $q^0(t)$:

$$\delta S(t_1,t_2)=\delta\int_{t_1}^{t_2} L(q,\dot{q},t)dt=\int_{t_1}^{t_2} L(q+\delta q,\dot{q}+\delta\dot{q},t)-L(q,\dot{q},t)dt \tag{2.33}$$

The difference in the right-hand side of Eq. 2.33 can be expressed through partial derivatives, so

$$\delta S(t_1,t_2)=\int_{t_1}^{t_2}\left(\frac{\partial L}{\partial q}\delta q+\frac{\partial L}{\partial \dot{q}}\delta\dot{q}\right)dt\,. \tag{2.34}$$

Taking into account the fact that $\delta\dot{q}(t)=\frac{d}{dt}\delta q(t)$ and integrating the second term of the integrand in the right-hand side of Eq. 2.34 by parts, we obtain the main expression for the variation of the action integral

$$\delta S(t_1,t_2)=\frac{\partial L}{\partial \dot{q}}\delta q\Bigg|_{t_1}^{t_2}+\int_{t_1}^{t_2}\left(\frac{\partial L}{\partial q}-\frac{d}{dt}\frac{\partial L}{\partial \dot{q}}\right)\delta q dt\,. \tag{2.35}$$

The first term here vanishes because the variation equals zero at the initial and final integration points t_1 and t_2. Finally, we obtain

$$\delta S(t_1,t_2)=\int_{t_1}^{t_2}\left(\frac{\partial L}{\partial q}-\frac{d}{dt}\frac{\partial L}{\partial \dot{q}}\right)\delta q dt\,. \tag{2.36}$$

The integral in the right-hand side of Eq. 2.36 equals zero for an arbitrary function $\delta q(t)$ if and only if the term in the parentheses identically equals zero. Hence, if a trajectory satisfies the equations of motion (Eq. 2.14) it also obeys the principle of least action (Eq. 2.32), and, vice versa, if the principle of least action (Eq. 2.32) is fulfilled for a trajectory, it must satisfy the Lagrangian equation of motion (Eq. 2.14).

2.2.3 Gauge Invariance of the Lagrangian

From the principle of least action it also immediately follows that the Lagrangian function is determined up to the total derivative with respect to time of any function of coordinates and time. In fact, if we add to the Lagrangian such a function $\frac{d}{dt}f(q,t)$, the action integral will change by only a constant $C=f(q_2,t_2)-f(q_1,t_1)$, which yields zero in the variation of the action integral (Eq. 2.31). Hence, the equation of motion will remain intact upon such an addition.

2.3 Nöther's Theorem and Conservation Laws

2.3.1 General Symmetries

The symmetry of in nature leads to some important consequences. There are general symmetries, namely the symmetry of space and time. It is assumed that space and time are homogeneous and isotropic. This means that the fundamental physical laws remain the same at any space point and at any instant of time and do not depend on the direction in space and time. Hence, physical laws in general and the equation of motion in particular must be invariant with respect to translations and rotations of space and time shift and reversal. This invariance with respect to transformations of space and time gives rise to the existence of very general conservation laws or constants of motion.

2.3.2 Time Invariance and the Energy Conservation Law

Consider first the invariance with respect to time displacements and reversal. The Lagrangian function does not depend explicitly on time and is invariant with respect to time reversal. The latter is true because the Lagrangian involves only quadratic terms in the time derivatives $\dot{q}_i$. Because the Lagrangian function does not depend explicitly on time, its total derivative with respect to time does not contain the partial derivative $\frac{\partial L}{\partial t}$, that is,

$$\frac{dL}{dt}=\sum_i \frac{\partial L}{\partial q_i}\dot{q}_i+\sum_i \frac{\partial L}{\partial \dot{q}_i}\ddot{q}_i\,. \tag{2.37}$$

The first term in Eq. 2.37 can be transformed according to the Lagrangian equations (Eq. 2.14) by substituting $\frac{d}{dt}\frac{\partial L}{\partial \dot{q}_i}$ for $\frac{\partial L}{\partial q_i}$, which gives

$$\frac{dL}{dt}=\sum_i \dot{q}_i\frac{d}{dt}\frac{\partial L}{\partial \dot{q}_i}+\sum_i \frac{\partial L}{\partial \dot{q}_i}\ddot{q}_i=\sum_i \frac{d}{dt}\left(\frac{\partial L}{\partial \dot{q}_i}\dot{q}_i\right). \tag{2.38}$$

Hence,

$$\frac{d}{dt}\left(\sum_i \frac{\partial L}{\partial \dot{q}_i}\dot{q}_i-L\right)=0\,. \tag{2.39}$$

Equation 2.39 means that the quantity

$$E = \sum_i \frac{\partial L}{\partial \dot{q}_i}\dot{q}_i - L \tag{2.40}$$

is a constant of motion. This quantity is called the energy of the system (or the total energy in order to distinguish it from various energy components). The energy conservation law is obeyed for any closed system and for a system in an external potential if this potential is independent of time. If we insert here the explicit expression for L, we readily obtain that

$$E = T + V. \tag{2.41}$$

Note here that $\sum_i \dot{q}_i \frac{\partial L}{\partial \dot{q}_i} = \sum_i \dot{q}_i \frac{\partial T}{\partial \dot{q}_i} = 2T$, first because V does not depend on $\dot{q}$, and then by virtue of Euler's theorem for homogeneous functions.

2.3.3 Additivity of Energy

Let a system can be subdivided into two (or more) subsystems such that $V = V_A + V_B$ and V_A and V_B depend only on the coordinates related to subsystems A and B, respectively. It is evident that in this case $L = L_A + L_B$ and $E = E_A + E_B$, where the corresponding terms depend only on the coordinates of the corresponding subsystems. We may say that subsystems A and B do not interact with each other. The subsystems may represent individual particles, so we are dealing with the system of noninteracting particles.

2.3.4 Spatial Invariance and Nöther's Theorem

Consider now the consequences of spatial invariance. Let, for example, the Lagrangian not depend on one of the generalized coordinates q_i. Such a coordinate is called the cyclic coordinate. Then, $\frac{\partial L}{\partial q_i} = 0$, and the Lagrangian equation (Eq. 2.14) for this coordinate is reduced to to $\frac{d}{dt}\frac{\partial L}{\partial \dot{q}_i} = 0$, that is, the quantity $\frac{\partial L}{\partial \dot{q}_i} = \text{const}$ is a constant of motion. According to Eq. 2.28, we may say that the momentum p_i conjugate to a cyclic generalized coordinate q_i is a conserved quantity.

In a more general case, consider a transformation of coordinates depending on one parameter α

$$q_i \to q'_i\,(\alpha) = A_i\,(q, \alpha), \tag{2.42}$$

which leaves the action integral of the system unchanged (q as usual designates the whole set of generalized coordinates). Let a trajectory $q(t)$ satisfy the Lagrangian equations for a system of particles and describe a mechanical state of the system. By virtue of the least action principle, the family of transformed trajectories

$$q'_i\,(t, \alpha) = A_i\,(q(t), \alpha) \tag{2.43}$$

generated by Eq. 2.42 will also satisfy the Lagrangian equations and, therefore, will also describe its possible mechanical states.

The invariance of the mechanical laws of motion means that transformation (Eq. 2.42) describes a certain symmetry transformation of the system. The famous Nöther's theorem[b] states that every symmetry transformation of a system generates a conserved quantity. Its proof for the particular case of coordinate transformations given by Eq. 2.42 is rather simple. The derivative of the action integral for the family of functions $q'_i(t, \alpha)$ with respect to the parameter α must equal zero, and we may write

$$\frac{d}{d\alpha}S(t_1, t_2, \alpha) = \int_{t_2}^{t_1}\sum_i\left(\frac{\partial L}{\partial q'_i}\frac{dq'_i}{d\alpha} + \frac{\partial L}{\partial \dot{q}'_i}\frac{d\dot{q}'_i}{d\alpha}\right)dt = 0. \tag{2.44}$$

Now assume that $\alpha \to 0$. Then, in the first order with respect to α $q'_i(t,\alpha) = q_i(t) + \alpha\dfrac{dA_i(q,\alpha)}{d\alpha}$ and hence, $\dot{q}'_i(t,\alpha) = \dot{q}_i(t) +,$ $\alpha\dfrac{d\dot{A}_i(q,\alpha)}{d\alpha}$ where the order of differentiation with respect to time and α can be changed because the transformation law does not depend on time. Finally, as α tends to zero, Eq. 2.44 transforms in the limit to

$$\frac{d}{d\alpha}S(t_1, t_2, \alpha)\Big|_{\alpha=0} = \int_{t_2}^{t_1}\sum_i\left(\frac{\partial L}{\partial q_i}\frac{dK_i}{d\alpha} + \frac{\partial L}{\partial \dot{q}_i}\frac{d\dot{A}_i}{d\alpha}\right)dt = 0. \tag{2.45}$$

Now, we integrate the second term in the parentheses by parts, proceeding in the same way as in the derivation of the principle of least action (see Eqs. 2.2–2.6) to obtain

[b]This very general theorem is applicable not only to mechanical but also to any physical system whose dynamical behavior is described by a variational principle.

$$\sum_i \left(\frac{\partial L}{\partial \dot{q}_i} \frac{dA_i}{d\alpha} \right) \Bigg|_{t_1}^{t_2} + \int_{t_1}^{t_2} \left(\sum_i \left(\frac{\partial L}{\partial q_i} - \frac{d}{dt} \frac{\partial L}{\partial \dot{q}_i} \right) \frac{dA_i}{d\alpha} \right) dt = 0 . \quad (2.46a)$$

For trajectories $q(t)$ that fulfill the Lagrangian equations (Eq. 2.14), the second term in the right-hand part of Eq. 2.46a equals zero, so

$$\sum_i \left(\frac{\partial L}{\partial \dot{q}_i} \frac{dA_i}{d\alpha} \right) \Bigg|_{t_1}^{t_2} = 0 . \quad (2.46b)$$

However, it follows from Eq. 2.46a that the values that the sum takes at two completely arbitrary instants of time must be equal. That is,

$$\sum_i \left(\frac{\partial L}{\partial \dot{q}_i} \frac{dA_i}{d\alpha} \right) = \text{const} , \quad (2.47)$$

which not only proves the theorem but also gives an explicit expression for the constant of motion associated with the particular symmetry transformation (Eq. 2.42).

2.3.5 Translational Invariance and the Momentum Conservation Law

Now, we consider the consequences of spatial invariance with respect to translations. Because these transformations are applied to each particle of the system, we revert here to Cartesian coordinates. The invariance with respect to translations means that the system is invariant with respect to a shift of each of its particles in space by an arbitrary vector $\alpha\mathbf{a}$ ($\mathbf{a}$ is a unit vector with an arbitrary direction). Hence, the symmetry transformation takes the form $\mathbf{r}_k \to \mathbf{r'}_k = \mathbf{r}_k + \alpha\mathbf{a}$, where the subscript k relates to an individual particle. Then, Eq. 2.47 can be written as

$$\sum_k \left(\frac{\partial L}{\partial x_k} a_x + \frac{\partial L}{\partial y_k} a_y + \frac{\partial L}{\partial z_k} a_z \right) = \text{const} . \quad (2.48)$$

However, the partial derivatives in the parentheses are just the Cartesian momentum components of the kth particle and the sum in the parentheses is the projection of the momentum vector of the kth particle $\mathbf{p}_k$ onto the direction $\mathbf{a}$. That is, $\sum_k (\mathbf{p}_k \mathbf{a}) = (\mathbf{P}\mathbf{a}) = \text{const}$, where the parentheses now designate the scalar product of two vectors and **P** is the total vector momentum of the system

$$\mathbf{P} = \sum_k \mathbf{p}_k = \sum_k m_k \mathbf{v}_k. \tag{2.49}$$

The constancy of the vector projection onto an arbitrary direction means the constancy of the vector itself. Hence, the total momentum of a closed system of particles is a conserved quantity and its three Cartesian components are constants of motion.

The conservation of the total momentum can also be interpreted as the existence of a specific point in the system of particles that moves with a constant velocity. This point is called the center of mass. Actually, we may define the conserved velocity of the entire system by dividing the total momentum by the total mass of the system $M = \sum_k m_k$:

$$\mathbf{V} = \frac{\mathbf{P}}{\sum_k m_k} = \frac{\sum_k m_k \mathbf{v}_k}{\sum_k m_k} = \frac{\sum_k m_k \dot{\mathbf{r}}_k}{\sum_k m_k} = \frac{d}{dt}\frac{\sum_k m_k \mathbf{r}_k}{\sum_k m_k} = \frac{d}{dt}\mathbf{R}^{\text{cm}} \tag{2.50}$$

It is evident that the point (center of mass) with the coordinates

$$\mathbf{R}^{\text{cm}} = \frac{\sum_k m_k \mathbf{r}_k}{\sum_k m_k} \tag{2.51}$$

moves with a constant velocity **V**.

2.3.6 Rotational Invariance and the Angular Momentum Conservation Law

The invariance with respect to rotations means that the system is invariant with respect to a rotation of all its particles simultaneously through an arbitrary angle around an arbitrary axis defined by a unit vector **a**. Again, it is convenient to use here the Cartesian coordinate system. Because we need only the derivative of the symmetry transformation with respect to its parameter, we consider a vector $\delta\alpha$ of an infinitesimal rotation around the unit vector **a** through an infinitesimal angle α, that is, $\delta\boldsymbol{\alpha} = \alpha\mathbf{a}$. The displacement of each particle of the system by this rotation is given by the vector product $\delta\mathbf{r} = [\alpha\boldsymbol{\alpha} \times \mathbf{r}] = \alpha[\boldsymbol{\alpha} \times \mathbf{r}]$. Hence, the symmetry transformation takes the form

$$\mathbf{r}_k \to \mathbf{r}'_k = \mathbf{r}_k + \delta\mathbf{r} = \mathbf{r}_k + \alpha[\boldsymbol{\alpha} \times \mathbf{r}]$$

In this particular case, Eq. 2.47 can be written as

$$\sum_k \left(\frac{\partial L}{\partial x_k}[\boldsymbol{\alpha} \times \mathbf{r}]_x + \frac{\partial L}{\partial y_k}[\boldsymbol{\alpha} \times \mathbf{r}]_y + \frac{\partial L}{\partial z_k}[\boldsymbol{\alpha} \times \mathbf{r}]_z \right) = \text{const}\,.$$

Inserting again the momentum components instead of the partial derivatives of the Lagrangian, rewriting this equation in vector form, and using the cyclic permutation of vectors in the scalar triple product, we finally obtain

$$\sum_k (\mathbf{p}_k[\boldsymbol{\alpha} \times \mathbf{r}_k]) = \sum_k (\boldsymbol{\alpha}[\mathbf{r}_k \times \mathbf{p}_k]) = \left(\boldsymbol{\alpha} \cdot \sum_k [\mathbf{r}_k \times \mathbf{p}_k] \right) = \text{const}\,.$$

We may again argue that the vector

$$\mathbf{L} = \sum_k [\mathbf{r}_k \times \mathbf{p}_k] \tag{2.52}$$

is a conserved quantity. This vector quantity is called the total angular momentum of the system. Note that both the momentum and the angular momentum are additive quantities.

2.3.7 Energy in a Laboratory Frame

In many cases, it is reasonable to introduce an inertial frame of reference in which the center of mass of an isolated system is at rest and consider only internal the motions of particles with respect to their center of mass. As distinct from the laboratory frame, in which the center of mass of a mechanical system under consideration can execute a motion (with a constant velocity), this frame is associated with the mechanical system and moves together with it. It is called the center-of-mass frame.

It is evident that the values of physical quantities in the laboratory and center-of-mass frames will differ. Thus, if in the laboratory frame the center of mass moves with a constant velocity $\mathbf{V}'$, the velocity of all the particles in the laboratory frame $\mathbf{v}_k^l$ will differ from their velocities in the center-of-mass frame $\mathbf{v}_k^{\mathrm{cm}}$ by the constant vector $\mathbf{V}'$ ($\mathbf{v}_k^l = \mathbf{v}_k^{\mathrm{cm}} + \mathbf{V}'$). Therefore, the kinetic energies calculated in both systems will also differ. Actually, we may write

$$T^l = \sum_k \frac{m_k (\mathbf{v}_k^l)^2}{2} = \sum_k \frac{m_k (\mathbf{v}_k^{\mathrm{cm}} + \mathbf{V}')^2}{2},$$

and removing the parentheses, we obtain

$$T^l = \frac{M(\mathbf{V}')^2}{2} + \mathbf{V}' \sum_k m_k \mathbf{v}_k^{\mathrm{cm}} + \sum_k \frac{m_k (\mathbf{v}_k^{\mathrm{cm}})^2}{2},$$

where $M = \sum_k m_k$ is the total mass of the system. However, the sum in the second term is the total momentum of the system in the center-of-mass frame, and hence $\sum_k m_k \mathbf{v}_k^{\mathrm{cm}} = \mathbf{P}^{\mathrm{cm}} = 0$ by the definition of the center of mass. Finally we obtain

$$T^l = T^{\mathrm{cm}} + \frac{M(\mathbf{V}')^2}{2} = T^{\mathrm{cm}} + \frac{(\mathbf{P}')^2}{2M}, \qquad (2.53)$$

where $\mathbf{P}' = M\mathbf{V}'$ is the total momentum of the system in the laboratory frame. T^{cm} may be considered as the internal kinetic energy of the system of particles, whereas the second term is the kinetic energy of the motion of the system in the laboratory frame considered as a whole (this term equals zero in the center-of-mass frame.

2.4 The Hamiltonian Function and Hamiltonian Equations

In many problems, it is convenient to use momenta rather than velocities as independent variables. To do that, we perform the so-called Legendre transformation. The total differential of the Lagrangian function is

$$dL = \sum_i \frac{\partial L}{\partial q_i} dq_i + \sum_i \frac{\partial L}{\partial \dot{q}_i} d\dot{q}_i,$$

which using Eqs. 2.14 and 2.15 can be written as

$$dL = \sum_i \dot{p}_i dq_i + \sum_i p d\dot{q}_i . \qquad (2.54)$$

The second term in Eq. 2.54 can be transformed to

$$\sum_i p_i d\dot{q}_i = d\left(\sum_i p_i \dot{q}_i\right) - \sum_i \dot{q}_i dp_i \,.$$

Transferring the total differential to the left-hand side and reversing the signs, we obtain

$$d\left(\sum_i p_i \dot{q}_i - L\right) = \sum_i \dot{q}_i dp_i - \sum_i \dot{p}_i dq_i \,.$$

The quantity in the parentheses is the total energy of the system. However, as independent variables, we consider now coordinates and momenta rather than coordinates and velocities. The total energy considered as a function of coordinates and momenta is called the Hamiltonian function and is given as

$$H(p,q) = \sum_i p_i \dot{q}_i - L = T(p) + V(q) \,.$$

From the equation for its total differential

$$dH(p,q,t) = \sum_i \dot{q}_i dp_i - \sum_i \dot{p}_i dq_i \,,$$

it immediately follows that

$$\frac{\partial H}{\partial p_i} = \dot{q}_i \,, \quad \frac{\partial H}{\partial q_i} = -\dot{p}_i \,. \qquad (2.55)$$

The system of Eq. 2.55 is a system of $2 \times 3N = 6N$ differential equations of the first order with respect to time for $6N$ unknown functions q_i, p_i. This is distinct from the Lagrangian (Newtonian) system, which is a system of $3N$ differential equations of the second order with respect to time for $3N$ unknown functions q_i (r_i).

2.4.1 Small Oscillations and Normal Coordinates

Now we consider one of the most important applications of classical mechanics for chemists. This is the problem of molecular vibrations. Another very important problem is *molecular dynamics*. Molecular dynamics will be considered in Chapter 4, Section 4.3. It is important here that both problems be treated within classical mechanics.

Consider first the potential energy function $V(q_i)$. Of our interest now is the case when this function has a minimum at some configuration $q_i = q_i^0$. This means that, first, all the partial derivatives of $V(q_i)$ equal zero $\frac{\partial V(q)}{\partial q_i} = 0$ for all $1 \leq i \leq 3N$ and, second, the matrix of the second derivatives with respect to q_i (the so-called Hessian matrix) is positive definite (for strict definitions, see Refs. [2, 3]). Such a configuration is an equilibrium point of the system because all forces vanish at this point. If the system assumes this configuration and all its velocities are zero, this means that it is at rest in this configuration.

Let us assume now that the particles in our system move in the vicinity of this configuration, so the amplitude of motion is sufficiently small. Then, the potential energy function can be expanded in a Taylor series, and retaining the first nonvanishing term of the second order yields

$$V(q_i) = V\left(q_i^0\right) + \frac{1}{2}\sum_{j,k} \frac{\partial^2 V}{\partial q_j \partial q_k}\left(q_j - q_j^0\right)\left(q_k - q_k^0\right).$$

It is reasonable to shift the origins of the energy scale and the frame, assuming that $V\left(q_i^0\right) = 0$ and $q_i^0 = 0$. Then, the potential energy takes the form $V(q_i) = \frac{1}{2}\sum_{ij} v_{ij} q_i q_j$, where we introduced the designation $v_{ij} = \frac{\partial^2 V}{\partial q_i \partial q_j}$, and, evidently, $v_{ij} = v_{ji}$. Such an approximation is called harmonic.

Using the expression for the kinetic energy in generalized coordinates (Eq. 2.17) and assuming that the coefficients $m_{ij}(q)$ equal their values[c] at q^0, $m_{ij}(q) = m_{ij}(q^0) = m_{ij}^0$, we obtain the equation of motion for the case of small oscillations in the form

$$\sum_j m_{ij}\ddot{q}_j + \sum_j v_{ij} q_j = 0. \tag{2.56}$$

Here, for convenience, we skipped the superscripts. The system of Eq. 2.56 is a system of linear homogeneous second-order differential equations.

[c]This is justified by our assumption that the oscillations and, hence, possible variations of coordinates are small.

Let us now select the generalized coordinates for our particular problem as the so-called mass-weighted Cartesian coordinates

$$q_i = \sqrt{M_i r_i} \ . \tag{2.57}$$

In these coordinates, the velocities can be written as $\dot{q}_i = \sqrt{M_i \dot{r}_i}$, and $m_{ij} = M_i\delta_{ij}$, where the Kronecker delta symbol $\delta_{ij} = 1$ for $i = j$ and $\delta_{ij} = 0$ for $i \neq j$. Hence, the kinetic energy simply becomes a sum of squared velocities $T = \frac{1}{2}\sum_i \dot{q}_i^2$. Hence, in these coordinates, Eq. 2.57 takes the form

$$\ddot{q}_i + \sum_j v_{ij} q_j = 0 , \tag{2.58}$$

where $v_{ij} = \frac{1}{\sqrt{M_i M_j}} \frac{\partial^2 V}{\partial r_i \partial r_j}$. The system of equations (Eq. 2.58) is a system of linear homogeneous second-order differential equations. It can be transformed by linear transformations to the form

$$\ddot{Q}_k + \lambda_k Q_k = 0 . \tag{2.59}$$

To do that, we define new coordinates $Q_k = \sum_i C_{ik} q_i$ and insert this expression into Eq. 2.59 to obtain

$$\sum_i C_{ik}\ddot{q}_i + \sum_i \lambda_k C_{ik} q_i = 0 . \tag{2.60}$$

It is evident now that we can cast Eq. 2.59 in the form of Eq. 2.60 by multiplying each of the equations in Eq. 2.59 by a coefficient C_{ik} and summing over index i[d]

$$\sum_i C_{ik}\ddot{q}_i + \sum_{ij} C_{ik} v_{ij} q_j = 0 . \tag{2.61}$$

Comparing Eqs. 2.60 and 2.61 yields immediately that the equality

$$\sum_{ij} C_{ik} k_{ij} q_j = \sum_i \lambda_k C_{ik} q_i \tag{2.62}$$

[d]What we are doing here is the diagonalization of the matrix $\boldsymbol{\lambda} = \left\| \lambda_{ij} \right\|$ by the unitary (orthogonal) matrix $\mathbf{C} = \left\| C_{ij} \right\|$. It is easy to show that $\mathbf{C}^{-1} = \mathbf{C}^T$ or in matrix elements, $\sum_k C_{ik} C_{kj} = \delta_{ij}$. By virtue of the aforesaid, if $Q_k = \sum_i C_{ik} q_i$, then $q_i = \sum_i C_{ik} Q_i$, for more details, see Refs. [2, 3].

must be fulfilled in order that Eq. 2.58 be true. Because coordinates q_i can take on any arbitrary values, we must assume that

$$\sum_j v_{ij} C_{jk} = \lambda_k C_{ik} . \tag{2.63}$$

(We changed the designation of dummy summation indices here and took into account the fact that $v_{ij} = v_{ji}$.) Equation 2.63 can be written as

$$\sum_j (v_{ij} - \lambda_k \delta_{ij}) C_{jk} = 0, \tag{2.64}$$

where we again use the Kronecker delta symbol to rearrange the equation.

Equation 2.64 has a nontrivial solution $C_{jk} \neq 0$ if and only if the determinant of this system vanishes, that is,

$$\left| v_{ij} - \lambda_k \delta_{ij} \right| = 0 . \tag{2.65}$$

This equation is a polynomial equation for λ_k of degree $3N$ and has $3N$ roots (some of which may coincide with each other). That is, we obtain $3N$ independent differential equations (Eq. 2.61) for functions $Q_i(t)$. Coordinates $Q_i(t)$ are called normal coordinates. Note that the kinetic and potential energies are simply expressed through normal coordinates

$$T = \frac{1}{2}\sum_i \dot{Q}_i^2 \text{ and } V = \frac{1}{2}\sum_i \lambda_i Q_i^2 , \tag{2.66}$$

that is, both functions are diagonal with respect to $\dot{Q}_i$ and Q_i. It is evident that $\frac{\partial^2 V}{dQ_i^2} = \lambda_i$. Hence, if the equilibrium point is stable, all λ_i are real and positive. In this case, each of the equations in Eq. 2.60 has the solution of the form

$$Q_i = A_i \cos\left(\sqrt{\lambda_i t + \varphi_i}\right), \tag{2.67}$$

where A_i and φ_i are arbitrary constants (amplitude and phase). For the Cartesian coordinates, we obtain

$$r_j = \frac{1}{\sqrt{M_j}} q_j = \frac{1}{\sqrt{M_j}} \sum_i C_{ji} Q_i = \frac{1}{\sqrt{M_j}} \sum_i C_{ji} A_i \cos\left(\sqrt{\lambda_i t + \varphi_i}\right). \tag{2.68}$$

If the system participates in the motion described by one normal mode, all the amplitudes A_i except one equal zero. In this case, each particle oscillates with the same frequency and phase, whereas the amplitude of its oscillations is proportional to its contribution to the particular normal coordinate and inversely proportional to the square root of its mass.

In conclusion, we must say that it turns out that all of this with only minor modifications is also applicable to the quantum-mechanical description of oscillations. Therefore, we will proceed to a very important chemical application of the theory, namely to molecular vibrations.

2.4.2 Classical Mechanics of Molecular Vibrations

The theory developed in the previous section will be applied now to the problem of molecular vibrations. Consider an isolated molecule composed of N atoms and assume that interactions that keep them together can be described with the use of a force field $V(r_i)$, where r_i are coordinates of atoms (or, more correctly, their nuclei) constituting the molecule. It is physically evident that the interaction energy, that is, the potential function must depend only on the mutual arrangement of atoms rather than on their absolute position in space. On the other hand, we know that the mechanical state of the molecule does not depend on the position of its center of mass in space and on the rotation of the molecule as a whole.

Three degrees of freedom always correspond to translational motion, and we can fix them, assuming that the total momentum of the molecule equals zero or, which is the same, assuming that the coordinates of the center of mass (Eq. 2.51) are constant. In terms of atomic displacements $\delta\mathbf{r}_k$ with respect to their equilibrium positions $\mathbf{r}_i^0$, this condition gives

$$\sum_k M_k \mathbf{r}_k = \text{const} = \sum_k M_k \mathbf{r}_k^0 .$$

Hence,

$$\sum_k M_k \delta\mathbf{r}_k = 0 . \tag{2.69}$$

In a general case, there are three rotational degrees of freedom (associated, for example, with three Eulerian [Euler] angles; see, for example, Refs. [2, 3]); however, if a molecule is linear, there are only two rotational degrees of freedom because the rotation around the axis aligned with the molecule does not affect the position of atoms in the linear molecule.

To fix the rotational degrees of freedom, it would be necessary to let the total angular momentum of the molecule (Eq. 2.52) be zero. However, this is, in general, impossible because the angular momentum is not the total derivative with respect to time of some function of coordinates. This means that strictly speaking, it is impossible to separate the rotational and vibrational degrees of freedom. Fortunately, this separation can be performed in the approximation of small oscillations. Consider again small variations $\delta \mathbf{r}_k$ of atomic positions with respect to their equilibrium positions $\mathbf{r}_i^0$. Then, the total angular momentum of the molecule will be written as

$$\mathbf{L} = \sum_k [\mathbf{r}_k \times \mathbf{p}_k] = \sum_k m_k [\mathbf{r}_k \times \mathbf{v}_k] = \sum_k m_k \left[\left(\mathbf{r}_k^0 + \delta \mathbf{r}_k \right) \times \delta \dot{\mathbf{r}}_k \right].$$

In the first approximation with respect to small variations $\delta \mathbf{r}_k$, this yields

$$\mathbf{L} \cong \frac{d}{dt} \sum_k m_k \left[\mathbf{r}_k^0 \times \delta \mathbf{r}_k \right].$$

The assumption that $\mathbf{L} = 0$ leads to the conclusion that

$$\sum_k m_k \left[\mathbf{r}_k^0 \times \delta \mathbf{r}_k \right] = \text{const} = 0, \qquad (2.70)$$

because this expression equals zero at $\delta \mathbf{r}_k = 0$.

Thus, a molecule has $3N - 6$ internal degrees of freedom for nonlinear equilibrium structures and $3N - 5$ internal degrees of freedom for linear equilibrium structures. In the simplest case of a diatomic molecule $A - B$ like N_2 or CO (linear case!), we have only one internal degree of freedom, which can naturally be chosen as its internuclear distance $R_{A\text{-}B}$. However, already in the three-atomic case, the situation becomes more complicated and the choice is no longer unique. Consider examples of water (H_2O) and ozone (O_3) molecules. In the former case, it is natural to choose two bond lengths ($R_{\text{O-H}_1}$ and $R_{\text{O-H}_2}$) and one bond angle (φ_{HOH}) as internal coordinates. In the second case, the choice is not unique, and the use of three internuclear distances might be more convenient for some problems.

2.5 Mechanical Similarity and the Virial Theorem

2.5.1 Homogeneity and Similarity

A function of n variables that obeys the condition

$$F(\lambda x_1, \lambda x_2, \ldots, \lambda x_n) = \lambda^k F(x_1, x_2, \ldots, x_n), \tag{2.71}$$

where λ is an arbitrary constant, is called the homogeneous function of variables x_i of degree k. For example, the kinetic energy is a homogeneous function of velocities of the second order. The potential energy function for small oscillations in the harmonic approximation problem is a homogeneous function of coordinates of the same order 2. The potential function of Coulomb interactions, which are the most important interactions in molecular theory, are also a homogeneous function of coordinates, but the degree of homogeneity equals −1 (they consist of terms of the form $\frac{Z_A Z_B}{\sqrt{(R_A - R_B)^2}}$).

Consider now a general case of a system whose potential function is a homogeneous function of order k. If we perform a transformation of coordinates $r_i \to \lambda r_i$, the potential energy will be multiplied by λ^k, whereas the kinetic energy will be multiplied by λ^2. To make both factors equal, we may additionally change the time scale and perform a transformation $t \to \mu t$. In this case, the kinetic energy will be multiplied by a factor of $\frac{\lambda^2}{\mu^2}$. Solving the equation $\frac{\lambda^2}{\mu^2} = \lambda^k$ gives the required value of the time scale factor $\mu = \lambda^{\frac{2-k}{2}}$.
In this case, the Lagrangian function will be multiplied by λ^k and the resulting equations of motion will remain unchanged.

This means that if $r_i(t)$ is a solution of the equations of motion, the function $r_i'(t) = \lambda r_i(\mu t)$ is also a solution of these equations. The trajectory $r_i'(t)$ is geometrically similar to $r_i(t)$ because these two curves are related by the similarity condition $\frac{r_i'(t')}{r_i(t)} = \lambda$, provided that two instants of time t' and t are in the ratio $\frac{t'}{t} = \mu = \lambda^{\frac{2-k}{2}} = \left(\frac{r_i'(t')}{r_i(t)}\right)^{\frac{2-k}{2}}$.

For small (and, actually, any harmonic) oscillations $\frac{2-k}{2}=0$, and we obtain that $\frac{t'}{t}=1$. Geometrically similar trajectories in this case correspond to oscillations with different amplitudes. Therefore, we may conclude that the period of small (or harmonic) oscillations does not depend on their amplitude.

2.5.2 The Virial Theorem

Systems with homogeneous potential functions of order k possess another important property. If the motion of a system is confined to some limited region of space, its average kinetic and potential energies $\overline{T}$ and $\overline{V}$ are related to each other by the simple equation

$$2\overline{T}=k\overline{V}\,, \tag{2.72}$$

This statement is known as the *virial theorem*. To prove it, we consider the sum $G=\sum_k \mathbf{p}_k\mathbf{r}_k$, where G is the virial of Clausius and the sum is taken over all particles of the system, and calculate its total derivative with respect to time as

$$\frac{d}{dt}\left(\sum_k \mathbf{p}_k\mathbf{r}_k\right)=\sum_k \mathbf{p}_k\dot{\mathbf{r}}_k+\sum_k \dot{\mathbf{p}}_k\mathbf{r}_k\,. \tag{2.73}$$

Note that $\mathbf{p}_k=\frac{\partial T}{d\mathbf{v}_k}$ by virtue of Eq. 2.24 and that $\sum_k \frac{dT}{d\mathbf{v}_k}\mathbf{v}_k=2T$ by virtue of the Euler's theorem on homogeneous functions. Then we obtain

$$\frac{d}{dt}\left(\sum_k \mathbf{p}_k\mathbf{r}_k\right)=2T+\sum_k \dot{\mathbf{p}}_k\mathbf{r}_k\,. \tag{2.74}$$

Now, we average this equation over time. By the average of a function $f(t)$ over time, we mean the quantity $\overline{f}=\lim_{\tau\to 0}\frac{1}{\tau}\int_0^\tau f(t)dt$.

After this averaging, the term in the left-hand side of Eq. 2.74 vanishes, and taking into account that $\dot{\mathbf{p}}_k=\frac{\partial V}{d\mathbf{r}_k}$ by virtue of the Newtonian equations, we obtain $2\overline{T}-\sum_{k}\mathbf{r}_k\overline{\frac{\partial V}{d\mathbf{r}_k}}=0$. Finally, considering that the potential function is a homogeneous function of order *k*, we finally obtain $2\overline{T}-k\overline{V}=0$, which proves the statement (Eq. 2.72) of the virial theorem.

For more details and further reading see Refs. [4, 5]. For some mathematical details, we also recommend Refs. [2, 3].

References

1. W. Heitler and F. London, Wechselwirkung neutraler Atome und homöopolare Bindung nach der Quantenmechanik, *Z. Phys.*, 1927, **44**, 455.
2. H. Eyring, J. Walter, and G. E. Kimball, *Quantum Chemistry*, John Wiley and Sons, New York, 1947.
3. V. Magnasco, *Elementary Molecular Quantum Mechanics: Mathematical Methods and Applications*, 2nd ed., Elsevier, Amsterdam, 2013, 932 p.
4. H. Goldstein, *Classical Mechanics*, 2nd ed., Addison-Wesley, Reading, 1980.
5. L. D. Landau and E. M. Lifshitz, *Mechanics*, 3rd ed., *Course of Theoretical Physics*, Vol. 1, Butterworth-Heinemann, 1976.

Chapter 3

Quantum Chemistry Methods for Molecular Disordered Materials

3.1 General Principles of Quantum Mechanics: A Brief Synopsis

3.1.1 Probabilistic Interpretation of Wave Function: The Variational Principle

There are many perfect books on quantum mechanics and its applications to molecular calculations [1–6]. Therefore, we will only briefly summarize here the main concepts, notions, and designations used in this book. We will stress the intimate connection between quantum and classical approaches, and even more so, as many molecular properties and characteristics are directly described in classical terms.

An analogue of classical trajectory in quantum mechanics is, generally speaking, the wave function $\Psi(q_i,t)$, which is determined in the space of coordinates of all particles of the system (configuration space) and depends on time. It determines the state of the system at each instant of time. The squared modulus of the wave function determines the probability density of finding the system at point $\mathbf{q} = \{q_i\}$ of the configuration space at time t. Hence $\int|\Psi(q_i)|^2 dq = 1$ (dq states for the product of all dq_i). A linear combination of two states $\Psi_1(q_i,t)$

Multiscale Modeling in Nanophotonics: Materials and Simulation
Alexander Bagaturyants and Mikhail Vener

ISBN 978-981-4774-40-6 (Hardcover), 978-1-315-10976-3 (eBook)
www.panstanford.com

and $\Psi_2(q_i,t)$, $\Psi_3(q_i,t) = c_1\Psi_1(q_i,t) + c_2\Psi_2(q_i,t)$, is also an admissible state (the principle of superposition). If $\int \Psi_1^*\Psi_2 dq = \int \Psi_2^*\Psi_1 dq = 0$, that is, functions Ψ_1 and Ψ_2 are orthogonal to each other, $|c_1|^2 + |c_2|^2 = 1$, and c_1 and c_2 are amplitudes of orthogonal (independent) states Ψ_1 and Ψ_2 in their superposition Ψ_3.

The wave function obeys the (time-dependent) Schrödinger equation (Eq. 3.1)

$$i\hbar\frac{\partial}{\partial t}\Psi = \hat{H}(\hat{q},\hat{p},t)\Psi\,, \tag{3.1}$$

where the Hamiltonian $\hat{H}(\hat{q},\hat{p},t) = \hat{T}(\hat{p}) + \hat{V}(\hat{q},t)$ is the total energy operator of the system in hand, $\hat{T}(\hat{p})$ is the kinetic energy operator, and $\hat{V}(\hat{q},t)$ is the potential energy operator. For a –conservative system $\hat{V}(\hat{q})$ is independent of t, and hence, operator $\hat{H} = \hat{H}(\hat{q},\hat{p})$ is also independent of t. The solutions of the eigenvalue problem (time-independent Schrödinger equation)

$$\hat{H}\Psi = E\Psi\,, \tag{3.2}$$

where eigenfunctions Ψ_i and eigenvalues E_i of $\hat{H}$ correspond to stationary states and energies of the system. Actually, from Eqs. 3.1 and 3.2, we directly obtain that $i\hbar\frac{\partial}{\partial t}\Psi_i = E_i\Psi_i$. Solving this equation gives $\Psi_i(q,t) = \exp(-i\frac{E_i}{\hbar}t)\Psi_i(q,t_0)$. The lowest eigenvalue of the Hamiltonian E_0 and the related eigenfunction Ψ_0 correspond to the ground state of the system such that the inequality

$$\langle\Psi|\hat{H}|\Psi\rangle \geq E_0 \tag{3.3}$$

holds for an arbitrary wave function, which is the essence of the *variational principle.* The variational principle is commonly used for the construction of approximate computational quantum-chemistry methods in the case when the approximate solution is constructed within a restricted class of trial functions.

3.1.2 Operators, Eigenvalues, and Eigenfunctions: The Projection Operator

A physical (observable) quantity A is described by a linear self-adjoint (Hermitian) operator $\hat{A}$, and its eigenvalues are real and correspond to values that quantity A can take on. The form of operator $\hat{A}$ is determined from the classical correspondence principle. In classical theory, any physical quantity A is a function of dynamical variables, and in the Hamiltonian formulation $A = f(q_i, p_i)$, where q_i and p_i are canonically conjugate coordinates and momenta. Substituting quantum-mechanical operators $\hat{q}_i$ and $\hat{p}_i$ for classical variables q_i and p_i into the classical expression for A, we obtain an operator expression for operator $\hat{A}$ in terms of operators $\hat{q}_i$ and $\hat{p}_i$. Here $\hat{q}_i$ is simply an operator of multiplication by q_i, while $\hat{p}_i = -i\hbar\frac{\partial}{\partial q_i}$.

Eigenfunctions and eigenvalues of operator $\hat{A}$ correspond to states $|a_i\rangle$ (eigenvectors) in which $\hat{A}$ obtains definite values a_i. Eigenvectors corresponding to different eigenvalues of a Hermitian operator are orthogonal to each other $\langle a_i | a_j\rangle = \delta_{ij}$. The expression of the form $|a\rangle\langle b|$ is an operator acting on the state vector as follows: $(|a\rangle\langle b|)|c\rangle = \langle b|c\rangle|a\rangle$. Consider an operator $\hat{P}_i = |a_i\rangle\langle a_i|$, where $|a_i\rangle$ are eigenvectors of operator $\hat{A}$. Acting by operator $\hat{P}_i$ on an arbitrary vector $|b\rangle$ gives $\hat{P}_i|b\rangle = |a_i\rangle\langle a_i|b\rangle$, where $\langle a_i|b\rangle$ are components of vector $|b\rangle$ in its expansion in vectors $|a_i\rangle$. Hence, $\hat{P}_i$ is the operator of projection onto the (1D) proper subspace corresponding to vector $|a_i\rangle$. It is easily verified that $\left(\hat{P}_i\right)^2 - \hat{P}_i = 0$ or $\left(\hat{P}_i - 1\right)\hat{P}_i = 0$. Note, also, that $\sum_i \hat{P}_i = I$, where I is the unit operator.

Operator $\hat{A}$ can be presented as its spectral decomposition $\hat{A} = \sum_i |a_i\rangle a_i \langle a_i|$. From this, it follows that $\hat{A}^n = \sum_i |a_i\rangle a_i^n \langle a_i|$. Generalizing this equation, it can be assumed as a definition that $f\left(\hat{A}\right) = \sum_i |a_i\rangle f(a_i)\langle a_i|$ for an arbitrary function. Let the operator equation $f\left(\hat{A}\right) = 0$ be obeyed. Then, a similar equation is obeyed for each eigenvalue $f(a_i) = 0$. Remember that any projection operator obeys the condition $\left(\hat{P}\right)^2 = \hat{P}$ or $\left(\hat{P} - 1\right)\hat{P} = 0$. Hence, the same equations are obeyed for its eigenvalues p_i, so $p_i = 0$ or 1.

Let us now construct an operator of projection $\hat{P}_1$ onto the proper subspace of operator $\hat{A}$ corresponding to its eigenvalue

a_1. It is easily verified that $\hat{P}_1 = \prod_{i \neq 1} \left(\frac{\hat{A} - a_i}{a_1 - a_i} \right)$. Actually, let $|b\rangle = \sum_i c_i |a_i\rangle$. Let us act onto vector $|b\rangle$ by operator $\hat{P}_1$. We obtain $\hat{P}_1 |b\rangle = \prod_{i \neq 1} \left(\frac{\hat{A} - a_i}{a_1 - a_i} \right) |b\rangle = \prod_{i \neq 1} \left(\frac{\hat{A} - a_i}{a_1 - a_i} \right) \sum_j c_j |a_j\rangle = c_1 |a_1\rangle$; that is, operator $\hat{P}_1$ actually projects an arbitrary vector $|b\rangle$ onto eigenvector vector $|a_1\rangle$ of operator $\hat{A}$.

3.1.3 The Virial Theorem

The virial theorem in quantum mechanics is a full analogue of this theorem in classical mechanics. Consider the commutator of the Hamiltonian $\hat{H} = V(\{X_i\}) + \sum_n \hat{P}_n^2 / 2m$ and the product $\hat{X}_n \hat{P}_n$ of the position operator $\hat{X}_n$ and the momentum operator $\hat{P}_n = -i\hbar \partial / \partial X_n$ of particle n $\left[\hat{H}, \hat{X}_n \hat{P}_n\right] = X_n \left[\hat{H}, \hat{P}_n\right] + \left[\hat{H}, \hat{X}_n\right] \hat{P}_n = i\hbar X_n \frac{\partial V}{\partial X_n} - i\hbar \frac{\hat{P}_n^2}{m}$.[a] Summing over all particles gives for $Q = \sum_n \hat{X}_n \hat{P}_n$ the commutator $\frac{i}{\hbar}[H, Q] = 2\hat{T} - \sum_n X_n \frac{\partial V}{\partial X_n}$ with the kinetic energy $\hat{T} = \sum_n \hat{P}_n^2 / 2m$. According to the Heisenberg equation of motion, the left-hand side of this equation is $-dQ/dt$. The expectation value of this time derivative vanishes in a stationary state, and we obtain the quantum virial theorem

$$2\langle \hat{T} \rangle = \sum_n \left\langle \hat{X}_n \frac{\partial V}{\partial \hat{X}_n} \right\rangle. \tag{3.4}$$

In many practically important cases, the potential V is a homogeneous function of coordinates, that is, $V(kX_1, \ldots, kX_n) = k^n f(X_1, \ldots, X_n)$. Thus, for harmonic vibrations, $k = 2$, while for Coulombic interactions, $k = -1$. For a homogeneous function, Euler's theorem is true, which states that $\sum_i x_i \frac{\partial f(x_i)}{\partial x_i} = kf(x_i)$. Hence, the virial theorem for homogeneous potential functions of degree k will read

[a]For the case of simplicity, we return here to Cartesian coordinates X_i.

as $2\langle\hat{T}\rangle = k\langle V\rangle$, and for Coulombic systems we have $2\langle\hat{T}\rangle = -\langle V\rangle$ or $\langle\hat{H}\rangle = \langle\hat{T}\rangle + \langle V\rangle = -\langle\hat{T}\rangle$. Equivalently, $\langle V\rangle = 2\langle\hat{H}\rangle$. These equalities strictly hold for exact solutions of the Schrödinger equation, while for approximate ones, they may be used to estimate the quality of approximations in the sense of balance between kinetic and potential energies.

3.1.4 The Hellmann–Feynman Theorem

An important consequence of the variational principle is the so-called Hellmann–Feynman theorem. In its most general formulation (generalized Hellmann-Feynman theorem), it states that if the Hamiltonian depends on a parameter λ, $\hat{H} = \hat{H}(\lambda)$, then the derivative of the Hamiltonian eigenvalue $E_n(\lambda)$with respect to λ, $\frac{\partial E_n}{\partial\lambda} = \frac{\partial}{\partial\lambda}\langle\Psi_n|\hat{H}|\Psi_n\rangle$, is expressed directly through the matrix element of the derivative of $\hat{H}$. That is,

$$\frac{\partial E_n}{\partial\lambda} = \langle\Psi_n|\frac{\partial\hat{H}}{\partial\lambda}|\Psi_n\rangle. \tag{3.5}$$

In fact, this equation was first derived by Pauli in 1933 [7], as was indicated and discussed by Musher [8]. Hans Hellmann derived this equation in his famous textbook, first published in Russian (Chapter 9, Paragraph 80, Eq. 80.5) and in the German edition (1937, Chapter 8, p. 285, Paragraph 54, Eq. 54.2). Independently, Feynman also derived this equality in 1939 [9]. Hellmann specially indicated the importance of Eq. 3.5 for molecular problems, when Ψ_n are electronic wave functions, and parameter λ is associated with nuclear coordinates.

3.1.5 Perturbation Theory: General Consideration

Sometimes, the Hamiltonian of a system, $\hat{H}$, only slightly differs from the Hamiltonian of another system, $\hat{H}^{(0)}$ (unperturbed or zero Hamiltonian), for which an exact solution is known. In this case, the perturbed Hamiltonian $\hat{H}$ can be presented as

$$\hat{H} = \hat{H}^{(0)} + \hat{V}, \tag{3.6}$$

where operator $\hat{V}$ is a perturbation. Then, $\hat{H}$ is a perturbed Hamiltonian, and $\hat{H}^{(0)}$ is an unperturbed or zero Hamiltonian. Now, we introduce a perturbation parameter λ into Eq. 3.6

$$\hat{H} = \hat{H}^{(0)} + \lambda\hat{V}\,, \tag{3.7}$$

such that $\hat{H} = \hat{H}^{(0)}$ when $\lambda = 0$, and $\hat{H} = \hat{H}^{(0)} + \hat{V}$ when $\lambda = 1$; that is, when $\lambda = 1$, Eq. 3.7 goes over into Eq. 3.6. The Schrödinger equation for the perturbed Hamiltonian takes the following form:

$$\hat{H}\left|\Psi_n\right\rangle = \left(\hat{H}^{(0)} + \lambda\hat{V}\right)\left|\Psi_n\right\rangle = E_n\left|\Psi_n\right\rangle. \tag{3.8}$$

Then, the solution of Eq. 3.8 can be expanded as a power series in parameter λ as

$$\left|\Psi_n\right\rangle = \left|\Psi_n^{(0)}\right\rangle + \lambda\left|\Psi_n^{(1)}\right\rangle + \lambda^2\left|\Psi_n^{(2)}\right\rangle + \dots \tag{3.9}$$

$$E_n = E_n^{(0)} + \lambda E_n^{(1)} + \lambda^2 E_n^{(2)} + \dots \tag{3.10}$$

Here, the following designations are used: Ψ_n is the nth perturbed eigenfunction, E_n is the nth perturbed eigenvalue, $\Psi_n^{(0)}$ is the nth unperturbed eigenfunction, $E_n^{(0)}$ is the nth unperturbed eigenvalue, $\Psi_n^{(k)}$ is the kth-order correction to the nth eigenfunction of the perturbed Hamiltonian, and $E_n^{(k)}$ is the kth-order correction to the nth eigenvalue of the perturbed Hamiltonian.

This is the so-called Rayleigh–Schrödinger perturbation theory. The set of unperturbed eigenfunctions obeys the following orthonormality conditions:

$$\left\langle\Psi_n^{(0)}\middle|\Psi_m^{(0)}\right\rangle = \delta_{n,m}\,.$$

It is convenient to introduce the so-called intermediate normalization condition for the perturbed eigenfunctions such that

$$\left\langle\Psi_n^{(0)}\middle|\Psi_n\right\rangle = 1 \tag{3.11}$$

Because of the normalization condition

$$\left\langle\Psi_n^{(0)}\middle|\Psi_n^{(0)}\right\rangle = 1\,,$$

from the intermediate normalization condition (Eq. 3.11), it follows that when $k > 0$,

$$\left\langle\Psi_n^{(0)}\middle|\Psi_n^{(k)}\right\rangle = 0\,. \tag{3.12}$$

Therefore, the intermediate normalization condition means that the only contribution made by the unperturbed eigenfunction Ψ^0 to the full wave function Ψ occurs in the zeroth order. (As a result, functions $\Psi_n^{(k)}$ obey the orthonormality condition $\left\langle \Psi_n^{(k)} \middle| \Psi_n^{(l)} \right\rangle = \delta_{k,l}$.)

Substituting Eqs. 3.9 and 3.10 into Eq. 3.8, we obtain

$$\left(\hat{H}^{(0)} + \lambda \hat{V}\right)\left(\left|\Psi_n^{(0)}\right\rangle + \lambda\left|\Psi_n^{(1)}\right\rangle + \lambda^2\left|\Psi_n^{(2)}\right\rangle + \ldots\right)$$
$$= \left(E_n^{(0)} + \lambda E_n^{(1)} + \lambda^2 E_n^{(2)} + \ldots\right)\left(\left|\Psi_n^{(0)}\right\rangle + \lambda\left|\Psi_n^{(1)}\right\rangle + \lambda^2\left|\Psi_n^{(2)}\right\rangle + \ldots\right). \tag{3.13}$$

Collecting together terms with the same degree of λ, we obtain the following equations for each degree of λ:

$$\hat{H}^{(0)}\left|\Psi_n^{(0)}\right\rangle = E_n^{(0)}\left|\Psi_n^{(0)}\right\rangle$$

$$\hat{H}^{(0)}\left|\Psi_n^{(1)}\right\rangle + \hat{V}\left|\Psi_n^{(0)}\right\rangle = E_n^{(0)}\left|\Psi_n^{(1)}\right\rangle + E_n^{(1)}\left|\Psi_n^{(0)}\right\rangle$$

$$\hat{H}^{(0)}\left|\Psi_n^{(2)}\right\rangle + \hat{V}\left|\Psi_n^{(1)}\right\rangle = E_n^{(0)}\left|\Psi_n^{(2)}\right\rangle + E_n^{(1)}\left|\Psi_n^{(1)}\right\rangle + E_n^{(2)}\left|\Psi_n^{(0)}\right\rangle \;\ldots$$

$$\hat{H}^{(0)}\left|\Psi_n^{(k)}\right\rangle + \hat{V}\left|\Psi_n^{(k-1)}\right\rangle = \sum_{i=0}^{k} E_n^{(i)}\left|\Psi_n^{(k-i)}\right\rangle. \tag{3.14}$$

Multiplying the obtained equations (Eq. 3.12) from the left by $\left\langle \Psi_n^{(0)} \right|$, we obtain corrections of the zero and the first order

$$\left\langle \Psi_n^{(0)} \middle| \hat{H}^{(0)} \middle| \Psi_n^{(0)} \right\rangle = \left\langle \Psi_n^{(0)} \middle| E_n^{(0)} \middle| \Psi_n^{(0)} \right\rangle = E_n^{(0)}\left\langle \Psi_n^{(0)} \middle| \Psi_n^{(0)} \right\rangle$$

$$\left\langle \Psi_n^{(0)} \middle| \hat{H}^{(0)} \middle| \Psi_n^{(1)} \right\rangle + \left\langle \Psi_n^{(0)} \middle| \hat{V} \middle| \Psi_n^{(0)} \right\rangle$$
$$= \left\langle \Psi_n^{(0)} \middle| E_n^{(0)} \middle| \Psi_n^{(1)} \right\rangle + \left\langle \Psi_n^{(0)} \middle| E_n^{(1)} \middle| \Psi_n^{(0)} \right\rangle$$

$$\left\langle \Psi_n^{(0)} \middle| \hat{H}^{(0)} \middle| \Psi_n^{(1)} \right\rangle + \left\langle \Psi_n^{(0)} \middle| \hat{V} \middle| \Psi_n^{(0)} \right\rangle$$
$$= E_n^{(0)}\left\langle \Psi_n^{(0)} \middle| \Psi_n^{(1)} \right\rangle + E_n^{(1)}\left\langle \Psi_n^{(0)} \middle| \Psi_n^{(0)} \right\rangle.$$

And with regard to Eqs. 3.11 and 3.12, we finally obtain

$$E_n^{(1)} = \left\langle \Psi_n^{(0)} \middle| \hat{V} \middle| \Psi_n^{(0)} \right\rangle. \tag{3.15}$$

From Eqs. 3.14 and 3.15, it follows that the energy of the nth state in the first order of perturbation theory,

$$E_n^1 = E_n^{(0)} + E_n^{(1)} = E_n^{(0)} + \left\langle \Psi_n^{(0)} \middle| \hat{V} \middle| \Psi_n^{(0)} \right\rangle$$
$$= \left\langle \Psi_n^{(0)} \middle| \hat{H}^0 \middle| \Psi_n^{(0)} \right\rangle + \left\langle \Psi_n^{(0)} \middle| \hat{V} \middle| \Psi_n^{(0)} \right\rangle$$
$$= \left\langle \Psi_n^{(0)} \middle| \hat{H}^0 + \hat{V} \middle| \Psi_n^{(0)} \right\rangle = \left\langle \Psi_n^{(0)} \middle| \hat{H} \middle| \Psi_n^{(0)} \right\rangle.$$

That is, the energy of the nth state in the first order of perturbation theory is just the average value of the total (perturbed) Hamiltonian over the nth eigenfunction of the zero Hamiltonian $\hat{H}^0$.

The second-order correction is obtained in a similar way:

$$\left\langle \Psi_n^{(0)} \middle| \hat{H}^{(0)} \middle| \Psi_n^{(2)} \right\rangle + \left\langle \Psi_n^{(0)} \middle| \hat{v} \middle| \Psi_n^{(1)} \right\rangle$$
$$= \left\langle \Psi_n^{(0)} \middle| E_n^{(0)} \middle| \Psi_n^{(2)} \right\rangle + \left\langle \Psi_n^{(0)} \middle| E_n^{(1)} \middle| \Psi_n^{(1)} \right\rangle + \left\langle \Psi_n^{(0)} \middle| E_n^{(2)} \middle| \Psi_n^{(0)} \right\rangle$$
$$\left\langle \Psi_n^{(0)} \middle| E_n^{(0)} \middle| \Psi_n^{(2)} \right\rangle + \left\langle \Psi_n^{(0)} \middle| \hat{v} \middle| \Psi_n^{(1)} \right\rangle$$
$$= E_n^{(0)} \left\langle \Psi_n^{(0)} \middle| \Psi_n^{(2)} \right\rangle + E_n^{(1)} \left\langle \Psi_n^{(0)} \middle| \Psi_n^{(1)} \right\rangle + E_n^{(2)} \left\langle \Psi_n^{(0)} \middle| \Psi_n^{(0)} \right\rangle$$

$$E_n^{(2)} = \left\langle \Psi_n^{(0)} \middle| \hat{V} \middle| \Psi_n^{(1)} \right\rangle \tag{3.16}$$

Finally, the correction of the kth order is given by the following equations:

$$\left\langle \Psi_n^{(0)} \middle| \hat{H}^{(0)} \middle| \Psi_n^{(k)} \right\rangle + \left\langle \Psi_n^{(0)} \middle| \hat{V} \middle| \Psi_n^{(k-1)} \right\rangle = \sum_{i=0}^{k} \left\langle \Psi_n^{(0)} \middle| E_n^{(i)} \middle| \Psi_n^{(k-i)} \right\rangle$$

$$\left\langle \Psi_n^{(0)} \middle| E_n^{(0)} \middle| \Psi_n^{(k)} \right\rangle + \left\langle \Psi_n^{(0)} \middle| \hat{V} \middle| \Psi_n^{(k-1)} \right\rangle = \sum_{i=0}^{k} E_n^{(i)} \delta_{0,k-i}$$

$$E_n^{(k)} = \left\langle \Psi_n^{(0)} \middle| \hat{V} \middle| \Psi_n^{(k-1)} \right\rangle \tag{3.17}$$

It is also useful to present here the mathematical formulation of the famous Wigner ($2n + 1$)-theorem [10], which states that the knowledge of $\Psi_n^{(k)}$ allows the calculation of the energy up to order $2n + 1$:

$$\begin{aligned} E_n^{(2k)} &= \left\langle \Psi_n^{(k)} \middle| \hat{V} \middle| \Psi_n^{(k-1)} \right\rangle - \sum_{i=1}^{k} \sum_{j=1}^{k} E_n^{(2k-i-j)} \left\langle \Psi_n^{(i)} \middle| \Psi_n^{(j)} \right\rangle \\ E_n^{(2k+1)} &= \left\langle \Psi_n^{(k)} \middle| \hat{V} \middle| \Psi_n^{(k)} \right\rangle - \sum_{i=1}^{k} \sum_{j=1}^{k} E_n^{(2k+1-i-j)} \left\langle \Psi_n^{(i)} \middle| \Psi_n^{(j)} \right\rangle \end{aligned} \tag{3.18}$$

Now, we will find perturbation corrections to the wave function. To do that, we present the kth-order correction to the perturbed wave function of the nth state as an expansion over the eigenfunctions of the zero-order (unperturbed) Hamiltonian

$$\left|\Psi_n^{(k)}\right\rangle = \sum_{\mu} C_{n,\mu}^{(k)} \left|\Psi_\mu^{(0)}\right\rangle. \tag{3.19}$$

We will again start from the second equation in Eq. 3.13, which determines the first-order perturbation corrections:

$$\sum_{\mu \neq n} C_{n,\mu}^{(1)} E_n^0 \left|\Psi_\mu^{(0)}\right\rangle + E_n^0 \left|\Psi_n^{(0)}\right\rangle = \sum_{\mu \neq n} C_{n,\mu}^{(1)} \hat{H}^{(0)} \left|\Psi_\mu^{(0)}\right\rangle + \hat{V} \left|\Psi_n^{(0)}\right\rangle$$
$$= \sum_{\mu \neq n} C_{n,\mu}^{(1)} E_\mu^0 \left|\Psi_\mu^{(0)}\right\rangle + \hat{V} \left|\Psi_n^{(0)}\right\rangle$$

To find the coefficients $C_{n,\mu}^{(1)}$, we multiply the perturbation equations by $\left|\Psi_\nu^{(0)}\right\rangle$, $\nu \neq n$, from the left.

$$\sum_{\mu \neq n} C_{n,\mu}^{(1)} E_n^{(0)} \left\langle \Psi_\nu^{(0)} \middle| \Psi_\mu^{(0)} \right\rangle + E_n^{(0)} \left\langle \Psi_\nu^{(0)} \middle| \Psi_n^{(0)} \right\rangle$$
$$= \sum_{\mu \neq n} C_{n,\mu}^{(1)} E_\mu^{(0)} \left\langle \Psi_\nu^{(0)} \middle| \Psi_\mu^{(0)} \right\rangle + \left\langle \Psi_\nu^{(0)} \middle| \hat{V} \middle| \Psi_n^{(0)} \right\rangle$$
$$C_{n,\nu}^{(1)} E_n^0 + 0 = C_{n,\nu}^{(1)} E_\nu^0 + \left\langle \Psi_\nu^{(0)} \middle| \hat{V} \middle| \Psi_n^{(0)} \right\rangle,$$

and hence,

$$C_{n,\nu}^{(1)} = \frac{\left\langle \Psi_\nu^{(0)} \middle| \hat{V} \middle| \Psi_n^{(0)} \right\rangle}{E_n^{(0)} - E_\nu^{(0)}} \tag{3.20}$$

so that

$$\left|\Psi_n^{(1)}\right\rangle = \sum_{\nu \neq n} C_{n,\nu}^{(1)} \left|\Psi_\nu^{(0)}\right\rangle = \sum_{\nu \neq n} \frac{\left\langle \Psi_\nu^{(0)} \middle| \hat{V} \middle| \Psi_n^{(0)} \right\rangle}{E_n^{(0)} - E_\nu^{(0)}} \left|\Psi_\nu^{(0)}\right\rangle \tag{3.21}$$

It is evident from Eqs. 3.20 and 3.21 that the unperturbed system must be nondegenerate. The case of degeneration will be considered separately.

Now, we can explicitly estimate not only the first-order correction to energy (which is given by Eq. 3.15) but also the second-order correction, using Eqs. 3.16 and 3.20. Substituting Eq. 3.20 into Eq. 3.16 yields

$$E_n^{(2)} = \left\langle \Psi_n^{(0)} \middle| \hat{V} \sum_{\nu \neq n} \frac{\left\langle \Psi_\nu^{(0)} \middle| \hat{V} \middle| \Psi_n^{(0)} \right\rangle}{E_n^{(0)} - E_\nu^{(0)}} \middle| \Psi_\nu^{(0)} \right\rangle,$$

and finally

$$E_n^{(2)} = \sum_{\nu \neq n} \frac{\left| \left\langle \Psi_\nu^{(0)} \middle| \hat{V} \middle| \Psi_n^{(0)} \right\rangle \right|^2}{E_n^{(0)} - E_\nu^{(0)}}. \tag{3.22}$$

Now, we will find second-order perturbation corrections to the wave function.

We will expand second-order corrections over the basis functions (eigenfunctions) of the zero-order Hamiltonian H_0

$$\left| \Psi_n^{(2)} \right\rangle = \sum_{\mu \neq n} C_{n,\mu}^{(2)} \left| \Psi_\mu^{(0)} \right\rangle, \tag{3.23}$$

and substitute expansions for both first-order and second-order corrections into the third equation in Eq. 3.13

$$\hat{H}^{(0)} \left| \Psi_n^{(2)} \right\rangle + \hat{V} \left| \Psi_n^{(1)} \right\rangle = E_n^{(0)} \left| \Psi_n^{(2)} \right\rangle + E_n^{(1)} \left| \Psi_n^{(1)} \right\rangle + E_n^{(2)} \left| \Psi_n^{(0)} \right\rangle.$$

Then, we obtain

$$\begin{aligned} &\sum_{\mu \neq n} C_{n,\mu}^{(2)} E_n^{(0)} \left| \Psi_\mu^{(0)} \right\rangle + \sum_{\mu \neq n} C_{n,\mu}^{(1)} E_n^{(1)} \left| \Psi_\mu^{(0)} \right\rangle + E_n^{(2)} \left| \Psi_n^{(0)} \right\rangle \\ &= \sum_{\mu \neq n} C_{n,\mu}^{(2)} \hat{H}^0 \left| \Psi_\mu^{(0)} \right\rangle + \sum_{\mu \neq n} C_{n,\mu}^{(1)} \hat{V} \left| \Psi_\mu^{(0)} \right\rangle \\ &= \sum_{\mu \neq n} C_{n,\mu}^{(2)} E_\mu^{(0)} \left| \Psi_\mu^{(0)} \right\rangle + \sum_{\mu \neq n} C_{n,\mu}^{(1)} \hat{V} \left| \Psi_\mu^{(0)} \right\rangle. \end{aligned} \tag{3.24}$$

To find the expansion coefficients, we again multiply perturbation equations (Eq. 3.19) from the left by $\left\langle \Psi_\nu^{(0)} \right|$, where $\nu \neq n$. This will result in

$$\begin{aligned} &\sum_{\mu \neq n} C_{n,\mu}^{(2)} E_n^{(0)} \delta_{\nu,\mu} + \sum_{\mu \neq n} C_{n,\mu}^{(1)} E_n^{(1)} \delta_{\nu,\mu} + E_n^{(2)} \delta_{\nu,n} \\ &= \sum_{\mu \neq n} C_{n,\mu}^{(2)} E^{(0)} \delta_{\nu,\mu} + \sum_{\mu \neq n} C_{n,\mu}^{(1)} \left\langle \Psi_\nu^{(0)} \middle| \hat{V} \middle| \Psi_\mu^{(0)} \right\rangle \end{aligned}$$

And, after summing the first three terms in this equation over index μ and taking into account that $\nu \neq n$, we obtain

$$C_{n,\nu}^{(2)} E_n^{(0)} + C_{n,\nu}^{(1)} E_n^{(1)} = C_{n,\nu}^{(2)} E_\nu^{(0)} + \sum_{\mu \neq n} C_{n,\mu}^{(1)} \left\langle \Psi_\nu^{(0)} \middle| \hat{V} \middle| \Psi_\mu^{(0)} \right\rangle.$$

From here, we finally obtain the wanted equation for coefficients $C_{n,\nu}^{(2)}$ of the second-order correction to the wave function

$$C_{n,\nu}^{(2)} = \frac{\sum_{\mu \neq n} C_{n,\mu}^{(1)} \left\langle \Psi_\nu^{(0)} \middle| \hat{V} \middle| \Psi_\mu^{(0)} \right\rangle - C_{n,\nu}^{(1)} E_n^{(1)}}{E_n^{(0)} - E_\nu^{(0)}}.$$

For higher-order corrections, we obtain

$$\sum_{i=0}^{k} \sum_{\mu \neq n} C_{n,\mu}^{(k-i)} E_n^{(i)} \left| \Psi_\mu^{(0)} \right\rangle$$
$$= \sum_{\mu \neq n} C_{n,\mu}^{(k)} \hat{H}^{(0)} \left| \Psi_\mu^{(0)} \right\rangle + \sum_{\mu \neq n} C_{n,\mu}^{(k-1)} \hat{V} \left| \Psi_\mu^{(0)} \right\rangle$$
$$= \sum_{\mu \neq n} C_{n,\mu}^{(k)} E_\mu^{(0)} \left| \Psi_\mu^{(0)} \right\rangle + \sum_{\mu \neq n} C_{n,\mu}^{(k-1)} \hat{V} \left| \Psi_\mu^{(0)} \right\rangle$$

Projecting again this equation onto the zero-order basis functions, we obtain

$$\sum_{i=0}^{k} \sum_{\mu \neq n} C_{n,\mu}^{(k-i)} E_n^{(i)} \delta_{\nu,\mu}$$
$$= \sum_{\mu \neq n} C_{n,\mu}^{(k)} E_n^{(0)} \delta_{\nu,\mu} + \sum_{\mu \neq n} C_{n,\mu}^{(k-1)} \left\langle \Psi_\nu^{(0)} \middle| \hat{V} \middle| \Psi_\mu^{(0)} \right\rangle,$$

Summing over the indices μ of the Kronecker deltas, we obtain

$$\sum_{i=0}^{k} C_{n,\nu}^{(k-i)} E_n^{(i)} = C_{n,\nu}^{(k)} E_\nu^{(0)} + \sum_{\mu \neq n} C_{n,\mu}^{(k-1)} \left\langle \Psi_\nu^{(0)} \middle| \hat{V} \middle| \Psi_\mu^{(0)} \right\rangle$$

And finally, we obtain an equation for $C_{n,\nu}^{(k)}$

$$C_{n,\nu}^{(k)} = \frac{\sum_{\mu \neq n} C_{n,\mu}^{(k-1)} \left\langle \Psi_\nu^{(0)} \middle| \hat{V} \middle| \Psi_\mu^{(0)} \right\rangle - \sum_{i=1}^{k} C_{n,\nu}^{(k-1)} E_n^{i}}{E_n^{(0)} - E_\nu^{(0)}},$$

and for $\left| \Psi_n^{(k)} \right\rangle$

$$\left| \Psi_n^{(k)} \right\rangle = \sum_{\mu \neq n} C_{n,\mu}^{(k)} \left| \Psi_\mu^{(0)} \right\rangle$$
$$= \frac{\sum_{\mu \neq n} C_{n,\mu}^{(k-1)} \left\langle \Psi_\nu^{(0)} \middle| \hat{V} \middle| \Psi_\mu^{(0)} \right\rangle - \sum_{i=1}^{k} C_{n,\nu}^{(k-i)} E_n^{(i)}}{E_n^{(0)} - E_\nu^{(0)}} \left| \Psi_\mu^{(0)} \right\rangle$$

Using these equations and the Wigner theorem, one can obtain corrections to energy in a compact form up to the fifth order.

Note here that the theory outlined before is applicable as is only to nondegenerate states, that is, to the case when $E_n^{(0)} - E_\nu^{(0)} \neq 0$. Moreover, parameter $\frac{\left\langle \Psi_\nu^{(0)} \middle| \hat{V} \middle| \Psi_n^{(0)} \right\rangle}{E_n^{(0)} - E_\nu^{(0)}}$ (see Eq. 3.20) indicates the extent to which the state $\left|\Psi_n^{(0)}\right\rangle$ is admixed to $\left|\Psi_\nu^{(0)}\right\rangle$ in the perturbed wave function due to perturbation $\hat{V}$.

3.1.6 Perturbation Theory: Degenerate States

Now we consider the case when the zero Hamiltonian $\hat{H}^0$ has some degenerate or quasi-degenerate levels, that is, when two or more of its eigenfunctions $\left|\Psi_k^{(0)}\right\rangle$ ($k = 1,\ldots, l$) belong to the same eigenvalue $E_a^{(0)}$ or their eigenvalues are rather close to each other. It is evident that any linear combination of functions $\left|\Psi_k^{(0)}\right\rangle$ is also an eigenfunction of $\hat{H}^0$ with the same eigenvalue $E_a^{(0)}$; that is, the set of functions $\left|\Psi_k^{(0)}\right\rangle$ ($k = 1,\ldots,l$) forms a linear subspace of dimensionality ι in the space of zero-order eigenfunctions $\left\{\left|\Psi_\mu^{(0)}\right\rangle\right\}$.

The problem is to find a solution of the equation

$$\hat{H}\left|\Psi\right\rangle = \left(\hat{H}^{(0)} + \lambda\hat{V}\right)\left|\Psi\right\rangle = E\left|\Psi\right\rangle, \tag{3.25}$$

such that its eigenvalue approaches the l-fold degenerate eigenvalue of the zero Hamiltonian when $\lambda \to 0$. Let us, for simplicity, designate the corresponding eigenfunctions of $\hat{H}^0$ as $\Psi_1^{(0)}$, $\Psi_2^{(0)}$, ... $\Psi_l^{(0)}$, so $E_1^0 = E_2^0 = \cdots = E_l^0$. We may consider that functions $\Psi_1^{(0)}$, $\Psi_2^{(0)}$, ... $\Psi_l^{(0)}$ are mutually orthogonal (because they can be made orthogonal by a linear transformation). As $\lambda \to 0$, Ψ must approach some solution of zero-order Hamiltonian whose eigenvalue equals E_1^0, that is, some linear combination

$$\Phi = c_1\Psi_1^0 + c_2\Psi_2^0 + \cdots + c_l\Psi_l^0, \tag{3.26}$$

where coefficients c_i are constants. This linear combination can be considered as a zero-order approximation to Ψ. The expansion of E and Ψ as a series in powers of λ will have the following form:

$$E = E_1^0 + \lambda E^{(1)} + \lambda^2 E^{(2)} \cdots, \qquad (3.27)$$

$$\Psi = \sum_{j=1}^{m} c_j \Psi_j^0 + \lambda \Psi^{(1)} + \lambda^2 \Psi^{(2)} + \cdots. \qquad (3.28)$$

Let us substitute these equations into Eq. 3.24 and take into account that coefficients at the same powers of λ must be equal. Then we obtain

$$\hat{H}^{(0)} \sum_{j=1}^{l} c_j \Psi_j^0 = E_1^0 \sum_{j=1}^{l} c_j \Psi_j^0 \qquad (3.29)$$

And

$$\left(\hat{H}^{(0)} - E_1^0\right)\Psi^{(1)} = \sum_{j=1}^{l} c_j \left(E^{(1)} - \hat{V}\right)\Psi_j^0. \qquad (3.30)$$

Because $E_1^0 = E_2^0 = \cdots = E_m^0$, Eq. 3.28 is satisfied. Now, as previously, we expand the unknown quantities in Eq. 3.29 as a linear combination of eigenfunctions of the zero-order Hamiltonian

$$\Psi^{(1)} = \sum_j A_j \Psi_j^0; \ \hat{V}\Psi_j^0 = \sum_k V_{kj} \Psi_k^0, \qquad (3.31)$$

where $V_{kj} = \left\langle \Psi_k^0 \middle| \hat{V} \middle| \Psi_j^0 \right\rangle$ and A_J are constants that should be determined. Then,

$$\sum_{j=1}^{l} c_j \hat{V} \Psi_j^0 = \sum_{j=1}^{l} \sum_k c_j V_{kj} \Psi_k^0 = \sum_{k=1}^{l} \sum_j c_k V_{jk} \Psi_j^0. \qquad (3.32)$$

Substituting these equations into Eq. 3.29 yields

$$\sum_{j=1}^{l} \left(E_j^0 - E_1^0\right) A_j \Psi_j^0 = \sum_{j=1}^{l} E^{(1)} c_j \Psi_j^0 - \sum_j \left(\sum_{k=1}^{l} c_k V_{jk} \Psi_j^0\right). \qquad (3.33)$$

The case of our interest is
$j \le l,\ E_j^0 = E_1^0$, so it follows from Eq. 3.32 that

$$\sum_{k=1}^{l} V_{jk} c_k - E^{(1)} c_j = 0. \qquad (3.34)$$

Equation 3.33 is a well-known secular equation, which can be presented in matrix form as $(\mathbf{V} - E^{(1)}\mathbf{I})\mathbf{c} = 0$, where **V** is the matrix of the perturbation operator $\hat{V}$, **I** is an identity matrix, and **c** is a vector of coefficients c_k.

It is well known from linear algebra that this equation has a nontrivial solution (i.e., a solution when $\mathbf{c} \neq 0$) only when the determinant of matrix $[\mathbf{V} - E^{(1)}\mathbf{I}]$ equals zero, that is, det $[\mathbf{V} - E^{(1)}\mathbf{I}] = 0$. This equation is an equation of the l th degree in $E^{(1)}$, and hence it has l roots $E_1^{(1)}$, $E_2^{(1)}$, ... $E_l^{(1)}$. Hence, there are m different (if only some of these roots are equal to each other) perturbed states whose energies approach E_1^0 as λ approaches zero.

Now, we may substitute a particular root $E_i^{(1)}$ for $E^{(1)}$ into the set of Eq. 3.33 and solve these equations for the ratios $\frac{c_2}{c_1}, \frac{c_3}{c_1}, \ldots$

Taking into account the normalization condition

$$c_1^* c_1 + c_2^* c_2 + \cdots + c_l^* c_l = 1,$$

one can determine all the coefficients c_i, and thus the zero-order eigenfunctions Φ_i, which subscript indicates the number of the root of Eq. 3.33 to which this particular eigenfunction corresponds.

Now, the first-order eigenfunction may be found by equating the coefficients of Ψ_j^0 for $j > l$ in Eq. 3.32. Then, we obtain

$$\left(E_j^0 - E_1^0\right) A_j = -\sum_{k=1}^{l} c_k V_{jk}, \text{ and } A_j = \frac{\sum_{k=1}^{l} c_k V_{jk}}{\left(E_1^0 - E_j^0\right)}.$$

Because functions Ψ must be normalized, it is necessary to put $A_j = 0$ for $j \leq l$. Hence, if Ψ_i is the eigenfunction for which the zero-order approximation is Φ_i, the first-order degenerate perturbation theory gives

$$\Psi_i = \Phi_i + \lambda \sum_{j>l} \frac{\sum_{k=1}^{l} c_k V_{jk}}{\left(E_1^0 - E_j^0\right)} \Psi_j^0 + \lambda^2(\cdots)$$

$$E_l = E_1^0 + \lambda E_l^{(1)} + \lambda^2(\cdots)$$

for the perturbed eigenfunctions and eigenvalues.

3.2 Molecular Hamiltonian and Wave Functions

3.2.1 Separation of Electronic and Nuclear Coordinates, Electronic and Nuclear Wave Equations, and Wave Functions

Molecules, as distinct from atoms, contain more than one atomic nucleus. Any molecular process somehow involves a nuclear rearrangement. Moreover, even in the molecular ground state the nuclei are in motion and execute small-amplitude oscillations around their equilibrium positions (zero-point oscillations).[b] To take into account nuclear motions, we should explicitly include nuclear coordinates in the molecular Hamiltonian and molecular wave function. Here and in the following presentation we do not consider the rotation of a molecule as a whole.[c] Then, the molecular Schrödinger equation will take the form[d]

$$\hat{H}(r,R)\Phi(r,R)=E^T\Phi(r,R), \qquad (3.35)$$

where r and R designate the sets of electronic and nuclear coordinates, respectively. E^T is the total energy of a molecule, and $\Phi(r, R)$ is the total wave function depending on both the electronic and nuclear coordinates, which obeys the normalization condition

$$\langle\Phi(r,R)|\Phi(r,R)\rangle_{r,R}=\int|\Phi(r,R)|^2\,drdR=1. \qquad (3.36)$$

The total Hamiltonian $\hat{H}(r,R)$ must include all the kinetic energy and potential energy terms as follows:

$$\hat{H}(r,R)=\hat{T}_{\mathrm{N}}(R)+\hat{V}_{\mathrm{NN}}(R)+\hat{T}_{\mathrm{e}}(r)+\hat{V}_{\mathrm{ee}}(r)+\hat{V}_{\mathrm{eN}}(r,R). \qquad (3.37)$$

Here, $\hat{T}_{\mathrm{N}}(R)$ is the nuclear kinetic energy operator

---

[b]Strictly speaking, in the case of an atom, the position of its nucleus cannot also be fixed. The condition that the total momentum of an atom equals zero means that its center of mass is at a fixed position. The nucleus position does not exactly coincide with the position of the center of mass, and the electrons and the nucleus of an atom execute collective motion around their common center of gravity.

[c]For more details, see Refs. [1–6] cited before.

[d]We use here (and throughout the book) the atomic unit system in which $\hbar=e=m_{\mathrm{e}}=1$, where $\hbar=\dfrac{h}{2\pi}$ is Planck's constant divided by 2π (sometimes called Dirac's constant), e is the elementary charge, and m_e is the electron mass.

$$\hat{T}_{\mathrm{N}}(R)=\sum_{A}\frac{\hat{p}_A^2}{2M_A}=-\sum_{A}\frac{1}{2M_A}\nabla_A^2, \tag{3.38}$$

and $\hat{V}_{\mathrm{NN}}(R)$ is the nuclear potential energy operator

$$\hat{V}_{\mathrm{NN}}(R)=\sum_{A>B}\frac{Z_A Z_B}{R_{AB}}, \tag{3.39}$$

where A, B designate nuclei, M_A and Z_A are their masses and charges, and R_{AB} are internuclear distances. $\hat{T}_{\mathrm{e}}(r)$ is the electronic kinetic energy operator

$$\hat{T}_{\mathrm{e}}(r)=\sum_{i}\frac{\hat{p}_i^2}{2}=-\sum_{i}\frac{1}{2}\nabla_i^2, \tag{3.40}$$

$\hat{V}_{\mathrm{ee}}(r)$ is the electron-electron interaction operator

$$\hat{V}_{\mathrm{ee}}(r)=\sum_{i>j}\frac{1}{r_{ij}}, \tag{3.41}$$

and $\hat{V}_{\mathrm{eN}}(r,R)$ describes the electron-nuclear interactions

$$\hat{V}_{\mathrm{eN}}(r,R)=-\sum_{A,i}\frac{Z_A}{r_{Aj}}. \tag{3.42}$$

Here, i, j designate electrons, and r_{ij} and r_{Aj} are electron-electron and electron-nucleus distances; we took into account that the electron charge equals −1 in the atomic unit system.

The molecular problem would be simplified if we could treat separately the nuclei and electrons, that is, *separate* the nuclear and electronic coordinates. The structure of Eqs. 3.35–3.42 indicates that the Hamiltonian $\hat{H}(r,R)$ contains a small term $\hat{T}_{\mathrm{N}}(R)$ proportional to $\frac{1}{M_A}\approx 0.00054$ and that the nuclear and electronic coordinates are mixed through the interaction term Eq. 3.42. Therefore, the entire system can be subdivided into two subsystems: the slow nuclear subsystem and the fast electronic subsystem. This means that a good starting point for the consideration of a molecular system might be an approximation in which we neglect the nuclear kinetic energy term Eq. 3.38 in Eq. 3.37. Actually, we may present the total Hamiltonian $\hat{H}(r,R)$ as the sum

$$\hat{H}(r\,|\,R)=\hat{T}_{\mathrm{N}}(R)+\hat{V}_{\mathrm{NN}}(R)+\hat{H}_{\mathrm{e}}(r\,|\,R), \tag{3.43}$$

where $\hat{H}_{\rm e}(r\,|\,R)$ might be considered as a purely electronic Hamiltonian for a system of electrons moving in the field of fixed nuclei:

$$\hat{H}_{\rm e}(r\,|\,R)=\hat{T}_{\rm e}(r)+\hat{V}_{\rm ee}(r)+\hat{V}_{\rm eN}(r,R)\,. \tag{3.44}$$

The bar in the parentheses separates the active electronic coordinates from the nuclear coordinates that serve as parameters of the electronic Hamiltonian.[e] This Hamiltonian defines wave equations (eigenvalue problem) for electronic wave functions

$$\hat{H}_{\rm e}(r\,|\,R)\Psi_{\rm e}(r\,|\,R)=E_{\rm e}(R)\Psi_{\rm e}(r\,|\,R) \tag{3.45}$$

in which nuclei are clamped at some fixed positions. Like the Hamiltonian, the electronic wave functions $\Psi_{\rm e}(r\mid R)$ depend on the nuclear coordinates parametrically, which is again indicated by the bar separating the active electronic coordinates r from the nuclear coordinates R, serving as parameters. We emphasize here that electronic wave functions $\Psi_{\rm e}(r\mid R)$ are defined in the space of electronic coordinates $\{r_i\}$ and are normalized as follows:

$$\left\langle \Psi_{\rm e}(r\,|\,R)\middle|\Psi_{\rm e}(r\,|\,R)\right\rangle_{\rm r}=\int\left|\Psi_{\rm e}(r\,|\,R)\right|^2 dr=1\,. \tag{3.46}$$

The electronic equation describes the motion of electrons in the field $\hat{V}_{\rm eN}(r,R)$ generated by nuclei fixed at positions R. It is evident that taking into account the normalization condition (Eq. 3.46),

---

[e]In the rigorous formulations, we should also include the so-called mass polarization term $\hat{H}_{\rm mp}(r)=-\frac{1}{2M}\left(\sum_i \nabla_i\right)^2$, where M is the total mass of the molecular system and the sum is taken over all electrons. This term arises from the transformation of molecular coordinates that is required to separate (i) the center of mass of the molecular system and (ii) the nuclear and electronic coordinates. To separate the nuclear and electronic coordinates, it is convenient to use the frame associated with the system of nuclei. This allows us to introduce internal coordinates of the nuclear frame in terms of bond lengths, bond angles, and dihedrals. However, a frame associated with the nuclear subsystem is not inertial with respect to a laboratory frame. In such coordinates, the center of mass of the entire molecular system (nuclei + electrons) does not move at a constant velocity. Hence, it cannot be fixed and executes a motion with respect to such a frame. The kinetic energy of this motion is added to the total Hamiltonian as the mass polarization term written before. This term is commonly neglected by most authors because it inversely proportional to the total mass and makes a relatively very small contribution to the total energy of the molecular system.

$E_e(R)$ can be presented as the average (observed) value

$$E_e(R) = \left\langle \Psi_e(r\,|\,R) \middle| \hat{H}_e \middle| \Psi_e(r\,|\,R) \right\rangle_r = \int \Psi_e^*(r\,|\,R) \hat{H}_e \Psi_e(r\,|\,R) dr\,. \quad (3.47)$$

Let us assume now that we solved electronic Eq. 3.45 and found the complete set of eigenstates $\Psi_{e,k}(r \mid R)$ and eigenvalues $E_{e,k}(R)$ of the electronic Hamiltonian $\hat{H}_e(r\,|\,R)$ at each nuclear configuration R. This basis set is called the adiabatic basis set. An electronic wave function $\Psi_{e,k}(r \mid R)$ and its energy $E_{e,k}(R)$ for particular k considered as functions of nuclear coordinates R determine what is called an adiabatic electronic state or an *electronic term*. Because, by definition, the basis set $\Psi_{e,k}(r \mid R)$ is complete at any point of the nuclear configuration space, we may expand a solution to the total molecular Schrödinger equation (Eq. 3.43) in the complete basis set $\Psi_{e,k}(r \mid R)$ to obtain (from here on, we will skip the subscript "e" because the designations are self-explanatory)

$$\Phi(r,R) = \sum_k \chi_k(R) \Psi_k(r\,|\,R), \quad (3.48)$$

where $\chi_k(R)$ are the coefficients of this expansion that must be determined at any point R. In fact, multiplying both sides of Eq. 3.48 by the (complex-conjugate) function $\Psi_k^*(r\,|\,R)$ and integrating over the electronic coordinates immediately yields

$$\chi_k(R) = \int \Psi_k^*(r\,|\,R) \Phi(r,R) dr\,, \quad (3.49)$$

that is, the functions $\chi_k(R)$ do exist, and any function $\Phi(r, R)$ can be presented in the form Eq. 3.48.

If we substitute Eq. 3.48 into Eq. 3.35, we obtain an equation for the nuclear functions $\chi_k(R)$ and the total energy E^T, which, in turn, determine the corresponding total molecular wave function $\Phi(r,R)$. This procedure of solving the Schrödinger equation using an expansion of the solution in terms of a convenient orthonormalized basis set of approximate wave functions is rather typical and is commonly used in many applications. Therefore, we will consider it now in more detail. The direct substitution of Eq. 3.48 into Eq. 3.35 yields

$$\hat{H}(r,R)\left(\sum_k \chi_k(R) \Psi_k(r\,|\,R) \right) = E^T \sum_k \chi_k(R) \Psi_k(r\,|\,R). \quad (3.50)$$

It is useful to recall here that $\Psi_k(r \mid R)$ are the eigenfunctions of the electronic Hamiltonian $\hat{H}_e(r|R)$ (see Eq. 3.45). With the use of Eq. 3.43, we easily obtain

$$\sum_k \hat{T}_N(R)\chi_k(R)\Psi_k(r|R) + \sum_k E_k(R)\chi_k(R)\Psi_k(r|R)$$
$$= E^T \sum_k \chi_k(R)\Psi_k(r|R). \tag{3.51}$$

Now, we may get rid of the electronic coordinates r and the electronic functions $\Psi_k(r \mid R)$ in Eq. 3.51 by passing on to matrix elements. To do that, we will multiply both sides of this equation by a certain function $\Psi_l(r|R)$ and integrate the obtained expressions over the electronic coordinates. Because the electronic eigenfunctions are orthonormalized in the space of electronic coordinates

$$\left\langle \Psi_{e,l}(r|R) \middle| \Psi_{e,k}(r|R) \right\rangle_r = \delta_{lk}. \tag{3.52}$$

we obtain a system of equations for the unknown functions $\chi_k(R)$ in the form

$$\sum_k \left\langle \Psi_l(r|R) \middle| \hat{T}_N(R)\chi_k(R)\Psi_k(r|R) \right\rangle_r + E_l(R)\chi_l(R)$$
$$= E^T \chi_l(R). \tag{3.53}$$

It is evident now that equations with different l in Eq. 3.53 are coupled through the first term associated with the nuclear kinetic energy. We will further transform this term so as to extract the nuclear kinetic energy operator $\hat{T}_N(R)$ from the matrix elements. By this means, it will be possible to isolate some small terms in the resulting equations that can be approximately neglected in the majority of chemically important cases.

To achieve this goal, we first simplify the representation of the nuclear kinetic energy operator $\hat{T}_N(R)$ by introducing the mass-weighted nuclear coordinates (as we did previously in classical mechanics; see Chapter 2, Eq. 3.72)

$$\mathbf{Q}_A = \sqrt{M_A}\mathbf{R}_A. \tag{3.54}$$

In these coordinates, the nuclear kinetic energy will take the form

$$\hat{T}_N(Q) = \tfrac{1}{2}\sum_i \hat{p}_{Q,i}^2 = -\tfrac{1}{2}\sum_i \frac{\partial^2}{\partial Q_i^2} = -\tfrac{1}{2}\nabla_Q^2, \tag{3.55}$$

where the designation ∇_Q stands for the multidimensional vector operator with components $\nabla_{Q,i} = \frac{\partial}{\partial Q_i}$. Now, we will expand the matrix element in the first term of Eq. 3.53 using this representation of the nuclear kinetic energy operator. We first consider the action of the operator (Eq. 3.2) on the product $\chi_k(Q)\Psi_k(r|Q)$. Double differentiation gives

$$\nabla_Q^2 \chi_k(Q)\Psi_k(r|Q) = \Psi_k(r|Q)\nabla_Q^2 \chi_k(Q)$$

$$+2(\nabla_Q \chi_k(Q)\cdot\nabla_Q \Psi_k(r|Q)) + \chi_k(R)\nabla_Q^2 \Psi_k(r|Q). \qquad (3.56)$$

Hence, we may write (to simplify the notation we will skip the designations of arguments in wave functions)

$$\hat{T}_N \chi_k \Psi_k = \Psi_k \hat{T}_N \chi_k - (\nabla_Q \chi_k \cdot \nabla_Q \Psi_k) - \tfrac{1}{2}\chi_k \nabla_Q^2 \Psi_k. \qquad (3.57)$$

Now, we have everything that we need to transform Eq. 3.35 to the final form. Substituting Eq. 3.57 into Eq. 3.51 gives (after some rearrangement of terms and taking into account the orthonormalization conditions, Eq. 3.46)

$$\hat{T}_N \chi_l + E_l(Q)\chi_l - \sum_k \left(\left\langle \Psi_l \middle| \nabla_Q \Psi_k \right\rangle \cdot \nabla_Q \chi_k\right)$$

$$-\tfrac{1}{2}\sum_k \left\langle \Psi_l \middle| \nabla_Q^2 \Psi_k \right\rangle \chi_k = E^T \chi_l. \qquad (3.58)$$

Equation 3.58 is a system of coupled differential equations for the unknown functions χ_l and the total molecular energy E^T. Of course, the solution of these equations is a formidable task that cannot be accomplished in practice, at least with present-day computational resources. Fortunately, the cross terms in Eq. 3.53 coupling functions χ_k and χ_l with $k \neq l$ are actually very small. These terms are proportional to matrix elements of derivatives of electronic wave functions with respect to nuclear coordinates. These derivatives are small because the electronic wave functions are slowly varying functions of nuclear coordinates. Therefore, a natural zero-order approximation called the *adiabatic approximation* is to neglect these cross terms, which is equivalent to retaining only one diagonal term in the sum given by Eq. 3.58

$$\Phi_l(r, Q) = \chi_l(Q)\Psi_l(r|Q). \qquad (3.59)$$

Therefore, in the adiabatic approximation, each electronic adiabatic state l gives rise to a set of nuclear functions χ_l and total molecular wave functions Φ_l directly associated with this electronic state. In the case of nondegenerate electronic states, which is most important for chemical applications, the diagonal matrix element in the third term of Eq. 3.58 vanishes,[f] and the equation for the nuclear function in the adiabatic approximation is reduced to the form

$$\hat{T}_{\mathrm{N}}\chi_l + E_l(Q)\chi_l - \tfrac{1}{2}\left\langle \Psi_l \middle| \nabla_Q^2 \Psi_l \right\rangle \chi_l = E^T \chi_l, \qquad (3.60)$$

which is determined by the related electronic adiabatic state. The first term on the left-hand side of this Hamiltonian is the usual nuclear kinetic energy operator. The second and the third terms represent operators of multiplication by some functions of nuclear coordinates Q and hence have the meaning of potential energy terms. Therefore, Eq. 3.60 can be interpreted as an effective Schrödinger equation for the nuclear wave function χ_l with an effective (adiabatic) Hamiltonian

$$\hat{H}_l^{\mathrm{ad}} = \hat{T}_{\mathrm{N}} + \hat{V}_l(Q)^{\mathrm{eff}}, \qquad (3.61)$$

where the effective potential $\hat{V}_l(Q)^{\mathrm{eff}}$

$$\hat{V}_l(Q)^{\mathrm{eff}} = E_l(Q) - \tfrac{1}{2}\left\langle \Psi_l \middle| \nabla_Q^2 \Psi_l \right\rangle, \qquad (3.62)$$

determines the motion of the nuclei. The first term in the potential Eq. 3.62 is the eigenenergy of the electronic Hamiltonian (which also includes the internuclear repulsion energy), that is, the energy of the electronic subsystem as a function of the nuclear configuration. The interpretation of the second term is somewhat more complicated. We note that the gradient of Ψ_l with respect to nuclear coordinates corresponds to the velocity at which the electronic wave function is dragged by the motion of nuclei. Hence, the third term may be considered as an additional contribution to the kinetic energy of electrons (averaged over the electronic wave function) due to their entrainment by the motion of nuclei. Recall that in the adiabatic approximation, we consider the motion of electrons in the field of

---

[f]The normalization integral $\langle \Psi_l | \Psi_l \rangle = 1$ is constant and does not depend on the nuclear coordinates R. Therefore, $\frac{\partial}{\partial R}\langle \Psi_l | \Psi_l \rangle = \left\langle \frac{\partial}{\partial R}\Psi_l \middle| \Psi_l \right\rangle + \left\langle \Psi_l \middle| \frac{\partial}{\partial R}\Psi_l \right\rangle = 0$. Hence, it follows that $\left\langle \Psi_l \middle| \frac{\partial}{\partial R}\Psi_l \right\rangle = 0$ if Ψ_l is real, and Ψ_l can always be made real for a nondegenerate eigenstate of a real (containing no complex terms) Hamiltonian operator.

the fixed nuclei. Therefore, the electronic Hamiltonian $\hat{H}_{\rm e}$ (Eq. 3.44) disregards this part of the electronic kinetic energy, and the third term in the effective Hamiltonian (Eq. 3.60) serves to compensate for this deficiency.

The next step in our approximate separation of nuclear and electronic motion is to neglect this small diagonal term as well. The approximation of this level is called the *Born–Oppenheimer (BO) approximation* by the names of the authors who developed this approach in 1927 [11]. It is this approximation that is most commonly used in chemical applications. The effective nuclear Hamiltonian in the BO approximation takes the form

$$\hat{H}_l^{\rm BO} = \hat{T}_{\rm N} + E_l(Q), \tag{3.63}$$

that is, in the BO approximation,

$$V_l(Q)^{\rm eff} = E_l(Q), \tag{3.64}$$

and the role of the nuclear potential energy is simply played by the energy of the electronic subsystem (which, in this case, also includes the energy of internuclear interaction). The BO approximation is a particular case of the adiabatic approximation in the sense that the wave function of the BO approximation is still presented in the form given by Eq. 3.59. Hence, the electronic wave functions and the electronic term $E_l(Q)$ in the BO approximation are the same as in the adiabatic approximation, and the two approximations differ only in the representation of the nuclear wave function.

In the problems that require a more accurate consideration, the nondiagonal (cross) terms in Eq. 3.58, which are neglected in the adiabatic and BO approximations, are commonly treated as nonadiabatic corrections perturbatively or in a variational calculation. The coefficients in these cross terms are matrix elements between the electronic wave functions of a given electronic term l and all other electronic terms k. A matrix element of this type can be estimated from perturbation theory[g] as

$$\left\langle \Psi_l \left| \frac{\partial}{\partial Q} \Psi_k \right. \right\rangle = \frac{\langle \Psi_l | \dfrac{\partial V_{\rm eN}}{\partial Q} | \Psi_k \rangle}{E_l(Q) - E_k(Q)}. \tag{3.65}$$

[g]See Chapter 3, Section 3.1.

Therefore, these matrix elements are small and the adiabatic approximation is applicable if the denominator in Eq. 3.65 is sufficiently large. That is, the adiabatic approximation is applicable in the regions where the given electronic term is sufficiently well separated in energy from other terms.

Sometimes, it is necessary to consider only a small area in the vicinity of a certain point R_0. In this case, the adiabatic approximation can be further simplified by using the solutions of the electronic equation Eq. 3.51 for the particular nuclear configuration R_0

$$\hat{H}_e(r\,|\,R_0)\Psi_e(r\,|\,R_0) = E_e(R_0)\Psi_e(r\,|\,R_0)\,. \tag{3.66}$$

This approximation is called the *crude Born–Oppenheimer (CBO) (or crude adiabatic) approximation.* We may again expand a solution to the total molecular Schrödinger equation in the series

$$\Phi(r,R) = \sum_k \chi_k(R)\Psi_k(r\,|\,R_0)\,. \tag{3.67}$$

Here, the electronic functions of the crude adiabatic approximation are the eigenfunctions of the electronic Hamiltonian at the fixed $R = R_0$

$$\hat{H}_e^{\text{CBO}}(r\,|\,R_0) = \hat{V}_{\text{NN}}(R_0) + \hat{T}_e(r) + \hat{V}_{\text{ee}}(r) + \hat{V}_{\text{eN}}(r,R_0)\,, \tag{3.68}$$

with the eigenvalues $E_k(R_0)$. The electronic wave functions still form a complete basis set in the space of functions of electronic coordinates. However, it must be noted that the convergence of this expansion becomes worse for nuclear configurations R such that $|R - R_0|$ is not small.

If we again single out the electronic Hamiltonian in the total molecular Hamiltonian, we obtain

$$\hat{H}(r,R) = \hat{T}_{\text{N}}(R) + \hat{H}_e^{\text{CBO}}(r\,|\,R_0) + \Delta V(r,R)\,, \tag{3.69}$$

where

$$\Delta V(r,R) = \hat{V}_{\text{NN}}(R) - \hat{V}_{\text{NN}}(R_0) + \hat{V}_{\text{eN}}(r,R) - \hat{V}_{\text{eN}}(r,R_0)\,. \tag{3.70}$$

Because the electronic wave functions in the CBO approximation no longer depend on the nuclear coordinates, there will be no matrix elements of the nuclear gradient between electronic wave functions in the total molecular Schrödinger equation similar to those in Eq. 3.60 of the adiabatic approximation. Instead, the analogous equation in the CBO approximation will contain matrix elements of the perturbation potential $\Delta V(r,R)$. The derivation of the equations for nuclear wave functions $\chi_k(R)$ is quite similar to that of Eq. 3.58, and we give here only the final result

$$\left[\hat{T}_{\mathrm{N}}+E_l\left(R_0\right)+\left\langle\Psi_l\left|\Delta V(r,R)\right|\Psi_l\right\rangle\right]\chi_l$$
$$+\sum_{k\neq l}\left\langle\Psi_l\left|\Delta V(r,R)\right|\Psi_k\right\rangle\chi_k=E^T\chi_l. \tag{3.71}$$

If we neglect the nondiagonal terms (collected in the sum in Eq. 2.71), we obtain the final equations of the CBO approximation

$$\left[\hat{T}_{\mathrm{N}}+E_l\left(R_0\right)+\left\langle\Psi_l\left|\Delta V(r,R)\right|\Psi_l\right\rangle\right]\chi_l=E^T\chi_l, \tag{3.72}$$

whereas the nondiagonal terms might be further treated as nonadiabatic corrections, if necessary.

As was mentioned before, the CBO approximation is applicable in a small vicinity of R_0. Therefore, in practice, $\Delta V(r,R)$ is commonly approximated by the linear term of the Taylor expansion

$$\Delta V(r,R)=\sum_i\frac{\partial V}{\partial R_i}\Delta R_i\ . \tag{3.73}$$

3.2.2 Adiabatic Terms and Potential Energy Surfaces: General Notions

In the expression for the effective nuclear Hamiltonian $H_l^{\mathrm{eff}}=\hat{T}_{\mathrm{N}}+V_l^{\mathrm{eff}}(Q)$, the effective nuclear potential energy function $V_l^{\mathrm{eff}}(Q)=E_l(Q)-\frac{1}{2}\left\langle\Psi_l\left|\nabla_Q^2\Psi_l\right.\right\rangle$ (Eq. 3.62) for the adiabatic approximation or $V_l^{\mathrm{eff}}(Q)=E_l(Q)$ (Eq. 3.64) for the BO approximation is often called the potential energy surface (PES). In fact, the function $V_l^{\mathrm{eff}}(Q)$ determines a surface in the space of $(M+1)$ coordinates (where M is the number of nuclear coordinates under consideration). If $M = 1$ (as, for example, in the case of a diatomic molecule), the PES becomes the potential energy curve (or simply, potential curve).

Solving the adiabatic equations for the electronic wave functions gives us a set (generally speaking, infinite, but finite in practice) of eigenvalues and eigenfunctions of the electronic Hamiltonian. Each of them generates its own PES for the nuclear equations, which again produce a set of nuclear eigenvalues and eigenfunctions. The local properties of a multidimensional surface are characterized by its derivatives.

For our purposes, it would suffice to consider the first derivatives, which form the gradient vector **g**(R) with components $g_i = \dfrac{\partial E(R_i)}{\partial R_i}$, and the second derivatives, which form a symmetric matrix $\mathbf{H} = \left[\dfrac{\partial^2 E(R_i)}{\partial R_i \partial R_j}\right]$ called the Hessian.[h] From the dynamical point of view, the gradient vector taken with the opposite sign has a meaning of the vector force acting on the nuclei **f = –g**(R), and the Hessian **H** represents the matrix of force constants. The shape of a PES and its general structure (topology) are in many instances determined by its critical (stationary) points, that is, points at which the gradient vector (or, in other words, the resultant force acting on the nuclei) equals zero

$$\mathbf{g}(R)\big|_{R=R'} = 0, \text{ or in components } g_i = \left.\frac{\partial E(R_i)}{\partial R_i}\right|_{R_i = R_i'} = 0\,. \quad (3.74)$$

Classically speaking, a stationary point corresponds to a stable or unstable equilibrium of the system. Local minima on the surface are stable equilibrium points of the system. At these points, the Hessian matrix **H** is positive definite, or, which is equivalent, all its eigenvalues are positive. Such points may be associated with some isomeric molecular structures that are relatively stable in a given electronic state. Hence, local minima on a PES correspond to various molecular isomers[i].

All other critical points of a PES can be further classified by the values of the second derivatives of at these points. The most important for chemical applications are critical points at which the matrix **H** has one negative eigenvalue with all others being positive. Such a critical point is called the first-order saddle point (or the true saddle point).

A PES can be picturesquely compared with a landscape with valleys (potential wells) surrounding minima on the surface and mountain passes (saddle points) connecting these valleys. Now, it is

---

[h]In mass-weighted coordinates, this matrix has the meaning of the force-constant matrix.

[i]A quantitative condition that must be satisfied in this case will be discussed later.

important for us that a minimum point is surrounded by the region (vicinity) in which the Hessian remains positive definite. This region might be associated with the potential well for a particular stable molecular structure. In most cases typical for molecular problems, we may approximately consider these potential wells separately, because solving the entire dynamical problem on a relatively complex multidimensional PES is a formidable task. From this point of view, a PES describes a set of stable molecular structures, which can interconvert to each other through corresponding saddle points.

Note here in conclusion that the forces acting on nuclei in a particular electronic state can be directly calculated from the corresponding electronic wave function and the nuclear-nuclear and electron-nuclear potential energy terms. Actually, according to the generalized Hellmann–Feynman theorem[j],

$$\frac{\partial E(R_i)}{\partial R_i} = \frac{\partial}{\partial R_i}\left\langle \Psi_e(r\,|\,R) \middle| \hat{H}_e \middle| \Psi_e(r\,|\,R) \right\rangle_r$$
$$= \left\langle \Psi_e(r\,|\,R) \middle| \frac{\partial}{\partial R_i}\hat{H}_e \middle| \Psi_e(r\,|\,R) \right\rangle_r. \quad (3.75)$$

Taking into account that only potential energy terms in the Hamiltonian $\hat{H}_e$ explicitly depend on nuclear coordinates, we finally obtain

$$\frac{\partial E(R_i)}{\partial R_i} = \left\langle \Psi_e(r\,|\,R) \middle| \frac{\partial \hat{V}}{\partial R_i} \middle| \Psi_e(r\,|\,R) \right\rangle_r, \quad (3.76)$$

where operator $\hat{V}$ is a sum of all potential energy terms in Eq. 3.44

$$\hat{V}(r,R) = \hat{V}_{NN}(R) + \hat{V}_{ee}(r) + \hat{V}_{eN}(r,R). \quad (3.77)$$

Equation 3.76 constitutes the content of the Hellmann–Feynman theorem in its conventional formulation. We see that operator $\hat{V}_{NN}(R)$ does not depend on electronic coordinates r and that operator $\hat{V}_{ee}(r)$ does not depend on nuclear coordinates R. Therefore, we may write

$$g_i(R) = \frac{\partial E(R_i)}{\partial R_i} = \frac{\partial \hat{V}_{NN}}{\partial R_i} + \left\langle \Psi_e(r\,|\,R) \middle| \frac{\partial \hat{V}_{eN}}{\partial R_i} \middle| \Psi_e(r\,|\,R) \right\rangle_r. \quad (3.78)$$

This result indicates that the gradient (and, hence, a quantum-mechanical force) is a purely one-electron property, so it can be

[j]See Section 3.1.

calculated using only the electron density without knowing the total many-electron wave function:

$$g_i(R) = \frac{\partial E(R_i)}{\partial R_i} = \frac{\partial \hat{V}_{\mathrm{NN}}}{\partial R_i} + \int \frac{\partial \hat{V}_{\mathrm{eN}}}{\partial R_i} \rho(r \mid R) dr \,. \tag{3.79}$$

Consider now the set of coordinates R_{Ai} of nucleus A. In this case, the first term in the right-hand side of Eq. 3.78 taken with the opposite sign represents the electrostatic force acting on nucleus A from all other nuclei of the system, and the second term taken with the opposite sign is the electrostatic force acting on nucleus A from the electrons averaged over the electronic wave function $\Psi_e(r \mid R)$ or, which is the same, over the electron density $\rho(r \mid R)$. This is the reason why the Hellmann–Feynman theorem is often called the electrostatic theorem. The Hellmann–Feynman electrostatic theorem demonstrates that all molecular interactions can be interpreted within classical electrostatic concepts and all the quantum effects are only due to the structure of the electronic wave function.

3.2.3 Nuclear Wave Equation and Wave Functions: General Quantum-Mechanical Treatment of Molecular Vibrations

In most important cases, the nuclear motion is restricted to a sufficiently small vicinity of the equilibrium structure of a certain isomer. This approximation is reasonable when the corresponding potential well is sufficiently deep and the temperature (i.e., the average kinetic energy per degree of freedom) is not too high. In this case, we may approximately consider the corresponding potential well separately from the entire surface as though it were a single minimum on the surface. The problem now is solved in a way similar to that used in classical mechanics[k].

In either the adiabatic or the BO approximation, the effective nuclear Hamiltonian can be written as

$$H_l^{\mathrm{eff}} = \hat{T}_{\mathrm{N}} + V_l(R), \tag{3.80}$$

where $V_l(R)$[l] for the given electronic term l depends only on nuclear coordinates R and may be considered the effective potential energy

---

[k]See Section 2.5.

[l]For convenience, we skipped here the superscript "eff."

for term l. The potential energy $V_l(R)$ can be expanded in the vicinity of the minimum point R_0 into a Taylor series up to the second-order terms. Passing to the normal coordinates Q defined in terms of displacements with respect to the equilibrium geometry R_0 (i.e., at $R = R_0$, $Q_0 = Q(R_0) = 0$; for more details, see Section 2.5, Eqs. 2.71–2.73), we will obtain the effective nuclear Hamiltonian in the form

$$H_l^{\text{eff}} = \tfrac{1}{2}\sum_i \frac{\partial^2}{\partial Q_i^2} + \tfrac{1}{2}\sum_i \lambda_i Q_i^2 = \tfrac{1}{2}\sum_i \left(\frac{\partial^2}{\partial Q_i^2} + \lambda_i Q_i^2 \right). \quad (3.81)$$

For simplicity the energy here is reckoned from the value $V_l(R_0) = E_l^0$. The Hamiltonian in Eq. 3.81 is a sum of independent Hamiltonians for 1D harmonic oscillators with force constants λ_i and frequencies $v_i = \sqrt{\lambda_i}$, so the problem of the multidimensional Hamiltonian given by Eq. 3.80 is reduced to $3N - 6$ independent equations for each normal coordinate ($3N - 5$ in the case of linear molecules).[m]

Hence, in the harmonic approximation, the nuclear wave function $\chi_l(R)$ in the total adiabatic molecular wave function (Eq. 3.59) is presented as a product of wave functions of 1D harmonic oscillators in the form

$$\chi_{lv}(Q) = \prod_i \chi_{lv_i}(Q_i), \quad (3.82)$$

where each oscillator wave function $\chi_{lv_i}(Q_i)$ corresponds to a certain normal mode of molecular vibrations, the subscript v_i indicates the vibrational quantum state of the ith mode, and the generalized index v designates the set of all the indices v_i ($v = \{v_i\}$). Finally, the total (adiabatic or BO) molecular wave function and energy will be presented in the form

$$\Phi_{lv}(r,Q) = \chi_{lv}(Q)\Psi_l(r\,|\,Q),$$

$$E_{lv}^T = E_{l0} + \sum_i \varepsilon_{iv_i}, \quad (3.83)$$

where ε_{iv_i} is the vibrational energy of the ith vibrational mode. According to the theory of linear harmonic oscillator (see, for example, Refs. [1–3])

---

[m]The remaining 6 (or 5 in the case of linear molecules) refer to the rotations of the molecule as a whole, which we do not consider here.

$$\varepsilon_{iv_i} = \left(n_i + \tfrac{1}{2}\right)\omega_i \,, \tag{3.84}$$

where n_i is the number of vibrational quanta in the vibrational state of the ith vibrational mode, and ω_i is the corresponding fundamental frequency.

The corresponding molecular electron vibrational states and energy levels are known as vibronic states and vibronic energy levels. At zero temperature $T = 0$ K, the system will reside in its vibrational ground state, that is, the state in which $n_i = 0$ for all vibrational modes i. That is, the total molecular energy for a particular local minimum in a particular electronic term in the ground vibronic state will be written as

$$E_{l0}^{T} = E_l^0\left(R_0\right) + \sum_i \tfrac{1}{2}\omega_i \tag{3.85}$$

and will differ from the minimum electronic energy $E_l^0\left(R_0\right)$ by the so-called zero-point energy (ZPE)

$$E^{\mathrm{ZPE}} = \sum_i \tfrac{1}{2}\omega_i \,. \tag{3.86}$$

The problem of vibronic interactions will be considered in more detail in a separate chapter.

3.2.4 Electronic Wave Equation and Wave Functions: One- and Two-Electron Operators

Let us now consider again the electronic wave equation

$$\hat{H}_{\mathrm{e}}(r\,|\,R)\Psi_{\mathrm{e}}(r\,|\,R) = E_{\mathrm{e}}(R)\Psi_{\mathrm{e}}(r\,|\,R)\,. \tag{3.87}$$

The electronic Hamiltonian can be presented as

$\hat{H}_{\mathrm{e}}(r\,|\,R) = \hat{T}_{\mathrm{e}}(r) + \hat{V}_{\mathrm{ee}}(r) + \hat{V}_{\mathrm{eN}}(r,R)$, where the kinetic energy operator $\hat{T}_{\mathrm{e}}(r) = -\sum_i \frac{1}{2}\nabla_i^2$, $\hat{V}_{\mathrm{ee}}(r) = \tfrac{1}{2}\sum_{i\neq j} r_{ij}^{-1}$, and $\hat{V}_{\mathrm{eN}}(r,R) = \sum_{in} \frac{Z_n}{\left|r_i - R_n\right|}$.

Consider now the average value of a one-electron operator

$$A_1 = \sum_{i=1}^{N} A(q_i)\,,\ \left\langle \Psi(1,2,\ldots,N)\right| A_1 \left|\Psi(1,2,\ldots,N)\right\rangle$$

$$= \sum_{i=1}^{N} \left\langle \Psi(1,2,\ldots i\ldots,N)\right| A(q_i) \left|\Psi(1,2,\ldots i\ldots,N)\right\rangle$$

$$= N\langle \Psi(1,2,\ldots i\ldots,N)|A(q_1)|\Psi(1,2,\ldots i\ldots,N)\rangle = \int A(1',1)\rho(1',1)d1'd1 .$$

If $A = V(q_1)$, then $A(1',1) = V(q_1)\delta(q_1 - q'_1)$, and finally we obtain $\rho_1(x \mid x') = N\int\Psi(x,x_2,\ldots,x_N)\Psi^*(x',x_2,\ldots,x_N)dx_2\ldots dx_N$. In a similar way, we can obtain

$$\rho_2(x_1x_2 \mid x_1'x_2')$$
$$= N(N-1)\int \Psi(x_1,x_2,x_3\ldots,x_N)\Psi^*(x_1',x_2',x_3\ldots,x_N)dx_3\ldots dx_N .$$

The electronic Hamiltonian contains one- and two-electron operators. In agreement with the variational principle, it is bounded below, and hence, at a given R, it has the lowest eigenvalue $E_{e0}(R)$ and a set of higher eigenenergies $E_{en}(R)$. These functions determine so-called electron terms. Many electronic properties are also expressed through one- and two-electron operators. Let us, first, consider a one-electron operator $\hat{f}(r,p)$. Because all electrons are identical, it should be presented as $\hat{f}(r,p) = \sum_i \hat{f}(r_i,\hat{p}_i)$. The average (observable) value of such an operator is

$$\langle \Psi_e(r)|\hat{f}(r,\hat{p})|\Psi_e(r)\rangle = \langle \Psi_e(r)|\sum_i \hat{f}(r_i,\hat{p}_i)|\Psi_e(r)\rangle$$

$$= N\int_{r_1} \hat{f}(r_1,\hat{p}_1)dr_1 \int_{r_2\ldots r_N} \Psi_e(r_1'r_2\ldots r_N)^*\Psi_e(r_1r_2\ldots r_N)dr_2\ldots dr_N .$$

The operator $\hat{f}(r,\hat{p})$ is determined by its kernel $F(r',r)$ so that

$$\hat{f}(r,\hat{p})\Psi_e(r) = \int_{r'} F(r',r)\Psi_e(r')dr' .$$

Here, we may define the first-order density matrix $\rho_1(r'|r) = N\int_{r_2\ldots r_N} \Psi_e(r'r_2\ldots r_N)^*\Psi_e(rr_2\ldots r_N)dr_2\ldots dr_N$, such that

$$\langle \Psi_e(r)|\hat{f}(r,p)|\Psi_e(r)\rangle = \langle \Psi_e(r)|\sum_i \hat{f}(r_i,\hat{p}_i)|\Psi_e(r)\rangle$$
$$= N\langle \Psi_e(r)|\hat{f}(r_1,\hat{p}_1)|\Psi_e(r)\rangle$$
$$= \int F(r',r)\rho_1(r'|r)dr'dr$$

In the case when operator $\hat{f}(r_1,\hat{p}_1)$ is simply a function of r_1, that is, $\hat{f}(r_1,\hat{p}_1) = f(r_1)$, its kernel $F(r',r) = F(r)\delta(r-r')$, and finally we have

$$\rho_1(r\,|\,r') = N\int \Psi(r,r_2,...,r_N)\Psi^*(r',r_2,...,r_N)dr_2...dr_N\,. \quad (3.88)$$

Similarly, we can consider a two-electron operator $\hat{f}(r',p';r'',p'') = \sum_{ij}\hat{f}(r_i,\hat{p}_i;r_j\hat{p}_j)$. We will not dwell here on further tedious mathematical transformations, but simply give the final result

$$\langle\Psi_e(r)|\hat{f}(r,p)|\Psi_e(r)\rangle = \langle\Psi_e(r)|\sum_{ij}\hat{f}(r_i,\hat{p}_i;r_j,\hat{p}_j)|\Psi_e(r)\rangle$$

$$= N(N-1)\int_{r_1r_2}\hat{f}(r_1,\hat{p}_1;r_2,\hat{p}_2)dr_1dr_2$$

$$\int_{r_3...r_N}\Psi_e(r_1r_2r_3...r_N)^*\Psi_e(r_1r_2r_3...r_N)dr_3...dr_N$$

$$= \int_{r_1r_2}\hat{f}(r_1,\hat{p}_1;r_2,\hat{p}_2)\rho_2(r_1r_2\,|\,r_1'r_2')dr_1dr_2\,.$$

Here, we introduced the so-called density matrices of the first and second orders,

$$\rho_1(r\,|\,r') = N\int \Psi(r,r_2,...,r_N)\Psi^*(r',r_2,...,r_N)dr_2...dr_N \quad (3.89)$$

and

$$\rho_2(r_1r_2\,|\,r_1'r_2') = N(N-1)\int \Psi(r_1,r_2,r_3...,r_N)\Psi^*(r_1',r_2',r_3'...,r_N')dr_2...dr_N$$

(3.90)

Using these density matrices, one can express average (observable) values of one- and two-electron operators without using wave functions explicitly.

3.2.5 Spin Operator and Its Properties

The total spin angular momentum operator ($\hat{S}$) for a system of N electrons and its square ($\hat{S}^2$) are defined as $\hat{S} = \sum_{i=1}^{N}\hat{S}_i$, $\hat{S}^2 = \hat{S}_x^{\,2} + \hat{S}_y^{\,2} + \hat{S}_z^{\,2}$. The components of $\hat{S}$ obey the commutator relations $\left[\hat{S}_a,\hat{S}_b\right] = i\hat{S}_c$, where a,b,c are x,y,z or any cyclic permutation of these indices. Now, we construct the operator of projection (see

Eq. 3.1) onto proper subspaces of the total spin squared operator $\hat{S}^2$. The eigenvalues of $\hat{S}^2$ are known well, $S_i^2 = S_i(S_i+1)$, so we obtain $\hat{P}_1 = \prod_{i \neq 1} \left(\frac{\hat{S}^2 - S_i(S_i+1)}{S_1(S_1+1) - S_i(S_i+1)} \right)$.

3.2.6 Ground and Excited Electronic States, Electronic Spectra, Energies, and Intensities of Electron Transitions

Among many electronic terms, there is one that reaches the lowest value $E_0(R)$ at some nuclear configuration R_0. This electronic term corresponds to the electronic ground state of the system $E_0(R)$, $\Psi_0(r \mid R)$, and $E_0(R)$ describes the ground-state PES. The other electronic terms located above the ground-state PES correspond to electronically excited states.

A chemist is most commonly interested in thermal chemical processes that occur on the ground-state PES. The structure and the topology of this surface determine many important properties of the particular molecular system at low energies and temperatures. On the other hand, many features and properties of the ground-state PES depend on the nature of the particular molecular system under consideration. The nuclear configuration R_0 corresponds to the global energy minimum on this surface. This global minimum commonly corresponds to the most stable structure of the system or to the most stable isomeric form if this system is a certain chemical molecule. This surface may also contain other minima that correspond to some other, less stable structures, which is almost always the case for polyatomic molecules.

Electronically excited states are very important in such areas as high-energy chemistry, radiation chemistry, photochemistry, and molecular spectroscopy. In many applications, low-lying excited states, less distant in energy from the ground state, are of most interest. Consider the equilibrium nuclear configuration R_0 on the ground-state PES and some low-lying (say, the lowest) excited state. In general, this configuration will not be a minimum for the PES of the excited state. However, we may still expect that the equilibrium molecular structure in a low-lying excited state will be close in geometry to the equilibrium ground-state structure. This means that the potential well of the excited state is located above the potential well on the ground-state PES, and the two terms can be approximated in the vicinity of their minima as two multidimensional paraboloids somewhat shifted with respect to one another.

A molecular system can pass from its ground state to an excited state and vice versa through, respectively, the absorption or emission of light. The energy of light quanta in this case should equal the excitation energy, that is, the energy difference between the excited and ground states, ΔE_{0i}. According to the theory of light-matter interaction, the spontaneous emission and induced absorption and emission intensities of a molecule are described by the Einstein coefficients. A molecule in the alternate electromagnetic field of a light wave will absorb or emit radiation and undergo quantum transitions. These transitions are called stimulated (or induced) transitions. The probabilities that a molecule will undergo such an induced transition per unit time from its lower state n to higher state m (induced absorption) $W^{a}_{n\to m}$, or vice versa, from an excited state m to a lower state n (induced emission) $W^{e}_{m\to n}$ are expressed via the Einstein coefficients $\boldsymbol{B}_{n\to m}$[n]

$$W^{a}_{n\to m} = B_{n\to m}u(\nu_{nm}) = W^{e}_{m\to n} = B_{m\to n}u(\nu_{nm}), \tag{3.91}$$

where $u(\nu_{nm})$ is the spectral radiant intensity and $\nu_{mn} = \dfrac{E_m - E_n}{h}$ is the frequency of the emitted light.

Even in the absence of an external electromagnetic field, a molecule in an excited state will spontaneously emit radiation and pass to a lower-lying state.[o] The total probability that a molecule will undergo a spontaneous transition from an upper state m to a lower state n per unit time is given by the Einstein coefficient $A_{m\to n}$. This coefficient is expressed via the coefficient $B_{m\to n}$

$$A_{m\to n} = \frac{(4\pi)^2 \hbar \nu^3_{mn}}{\mathbf{c}^3} B_{m\to n}, \tag{3.92}$$

[n]We consider for simplicity the case of nondegenerate states. Otherwise, degeneration factors g_m and g_n should be taken into account so that $g_n B_{n\to m} = g_m B_{m\to n}$.

[o]Spontaneous emission is a purely relativistic effect. In nonrelativistic quantum mechanics, any stationary state will exist forever unless it interacts with an external field. The reason for spontaneous emission in relativistic quantum theory is the interaction of the electron system with vacuum fluctuations of the electromagnetic field.

where **c** is the velocity of light. In spectroscopy, the frequency is commonly expressed in wave numbers $\tilde{v}_{nm} = \frac{v_{nm}}{c}$. Then Eq. 3.92 takes the form

$$A_{m\to n} = (4\pi)^2 \hbar \tilde{v}_{mn}^3 B_{m\to n}. \tag{3.93}$$

From the theory of light-matter interaction in the so-called dipole approximation, it follows that

$$B_{n\to m} = B_{m\to n} = \frac{8\pi^3}{3h^2}\left|\mathbf{d}_{nm}\right|^2 = \frac{2\pi}{3\hbar^2}\left|\mathbf{d}_{nm}\right|^2, \tag{3.94}$$

where $\mathbf{d}_{nm}$ is the matrix element of the electric dipole moment (electric dipole transition moment)

$$\mathbf{d}_{nm} = \left\langle \Psi_n \middle| \mathbf{d} \middle| \Psi_m \right\rangle = \left\langle \Psi_n \middle| \sum_k Z_k \mathbf{R}_k + e\sum_i \mathbf{r}_i \middle| \Psi_m \right\rangle, \tag{3.95}$$

Z_k is the charge of the kth nucleus, e is the electronic charge, and $\mathbf{R}_k$ and $\mathbf{r}_i$ are the radius vectors of the kth nucleus and the ith electron, respectively. The corresponding equation for $A_{m\to n}$ is obtained from Eqs. 3.93 and 3.54, and it follows that

$$A_{m\to n} = \frac{32\pi^3 \tilde{v}_{mn}^3}{3\hbar}\left|\mathbf{d}_{nm}\right|^2. \tag{3.96}$$

It is convenient to express the intensities of light absorption in terms of dimensionless values called (absorption) *oscillator strengths*, f, which are defined as

$$f_{nm} = \frac{mhv_{mn}}{\pi e^2} B_{n\to m} = \frac{8\pi^2 m v_{mn}}{3he^2}\left|\mathbf{d}_{nm}\right|^2. \tag{3.97}$$

Oscillator strengths are directly related to the (integral) intensities of absorption lines that are observed experimentally, so

$$f_{nm} = 2303\frac{m\mathbf{c}^2}{N\pi e^2 n}\int \varepsilon(\tilde{v})d\tilde{v} = \frac{4.32\times 10^{-9}}{n}\int \varepsilon(\tilde{v})d\tilde{v}. \tag{3.98}$$

Here, m and e are the mass and charge of an electron, respectively; **c** is the speed of light; n is the index of refraction; $\tilde{v}$ is the frequency of light expressed in wave numbers; and f is normalized to unity (i.e., $f_{max} = 1$) [12].

3.2.7 Spin-Orbit Interaction

Spin-orbit (SO) interaction is a purely relativistic term. However, its effect is important, even for low-energy electronic transitions, because it determines the possibility of electron transitions between states with different spin multiplicities, which are formally forbidden in nonrelativistic theory (see for more details Refs. [13–14]). "Its impact on molecular properties increases with nuclear charge Z to an extent that molecules containing heavy elements cannot be described correctly even in a qualitative manner, if spin-orbit coupling is not taken into account" [14]. SO interaction is described by the Breit–Pauli SO Hamiltonian, H^{SO}, which includes both one-electron and two-electron terms. However, two-electron terms can be approximately taken into account in an effective one-electron SO term. In this approach, the effective nuclear charge (Z_{eff}) approximation is the simplest method available for the qualitative computation of SOC effects in molecules. The Z_{eff} approximation consists of using only the one-electron term from the Breit–Pauli Hamiltonian

$$\hat{H}^{SO} = \frac{e^2}{2m^2c^2}\sum_{i=1}^{Nel}\sum_{\alpha=1}^{Nat}\frac{Z_{\text{eff}}(\alpha)}{r_{i\alpha}^3}\mathbf{l}_{i\alpha}\cdot\mathbf{s}_i\,, \tag{3.99}$$

where the nuclear charges $Z_{\text{eff}}(\alpha)$ are empirical parameters, which are determined by fitting to experimental atomic SO splitting [15]. In Eq. 3.99, $\mathbf{l}_{i\alpha}$ and $\mathbf{S}_i$ are space and spin angular momentum operators, $\hat{\mathbf{l}}_{i\alpha} = (\mathbf{r}_i - \mathbf{R}_\alpha)\times\hat{\mathbf{p}}_i$ and $\hat{\mathbf{l}}_{ij} = (\mathbf{r}_i - \mathbf{r}_j)\times\hat{\mathbf{p}}_i$. Higher-order SO effects can be determined by quasi-degenerate perturbation theory (see Section 3.1.6). In this approach, an interaction matrix is set up including all electrostatic and SO coupling matrix elements between the states under consideration. For more details, see Refs. [13–15].

3.3 MO LCAO Treatment of Molecular Systems

3.3.1 Hydrogen Molecular Ion and the MO LCAO Approximation

The simplest diatomic system is the hydrogen molecular ion H_2^+ (see Fig. 3.1). It contains two hydrogen nuclei and one electron. Just as the hydrogen atom has been considered as a prototypical system for other atoms, the hydrogen molecular ion may serve as a prototypical system for diatomic molecules. Many general properties of electronic wave functions in diatomic systems can be understood from studying the properties of solutions for H_2^+. Though the problem of electronic states in H_2^+ allows an exact solution in prolate spheroidal coordinates, this solution is not transparent. Instead, we consider a simple way of constructing an approximate molecular one-electron wave function as a linear combination of atomic one-electron wave functions. Molecular one-electron wave functions are generally called molecular orbitals (MOs). Atomic one-electron wave functions (atomic orbitals, AOs) are considered in many textbooks (see, for example, Refs. [2–6]). Therefore, this approximation is called MO LCAO (where the abbreviation LCAO stands for *linear combination of atomic orbitals*).

$$[\mathrm{H}_1\text{–}\mathrm{H}_2]^+$$

Figure 3.1 Hydrogen molecular ion H_2^+.

The fundamental principles of this approximation were developed by F. Hund and R. S. Mulliken at the dawn of the development of quantum chemistry (a historical outline of this revolutionary development is given in Ref. [15]). The main idea is that the molecular field in the vicinity of an atomic nucleus is close to the field in the corresponding atom. Hence, at least in the vicinity of an atomic nucleus, MOs will be described well by LCAOs of the corresponding atom.

We will start with the simplest approach in which the electronic wave function of H_2^+ is approximated as a linear combination of the atomic ground-state (1s) orbitals of hydrogen atoms constituting the H_2^+ molecular ion. Hence, this approximation can be written as follows:

$$\psi(r,s)=\psi(r)\sigma(s)=\left(c_1\varphi_1(r)+c_2\varphi_2(r)\right)\sigma(s), \qquad (3.100)$$

where $\varphi_1(r)$ and $\varphi_2(r)$ are 1s AOs of the first and the second hydrogen atoms, r and s are the spatial and spin electron coordinates, and c_1 and

c_2 are some arbitrary coefficients. In Eq. 3.100, we used the fact that the electron wave function in the absence of magnetic interactions can be factorized into the spatial and spin parts. Note here that $\psi(r,s)$ is called *spin orbital*, while $\psi(r)$ is called *spatial orbital* (which is normalized in the space of electronic coordinates), and $\sigma(s)$ is the *spin function*. The spin function $\sigma(s)$ can take only two values, which correspond to two possible projections of the electron spin onto a certain axis (axis z for definiteness), spin-up (α) and spin-down (β). Without any loss of generality, we can assume that the electron in our system has spin-up; that is, the projection of spin onto the z axis equals +(1/2) and $\sigma(s) = \alpha$.

In an explicit form, the (normalized) functions $\varphi_1(r)$ and $\varphi_2(r)$ can be written as

$$\varphi_i(r) = \frac{1}{\sqrt{\pi}} e^{-(|r-R_i|)}. \quad (3.101)$$

Coefficients c_1 and c_2 are unknown. We will find them using the variational principle, that is, we will select c_1 and c_2 in such a way that the electron energy for the wave function described by Eq. 3.100 is a minimum. In this case, the function found in this way will be the closest in energy to the exact solution of the Schrödinger equation for the hydrogen molecular ion and will provide an upper estimate for the exact energy.

Consider now the electronic Hamiltonian[p] operator (Eq. 3.44) for H_2^+, assuming that the hydrogen nuclei 1 and 2 are located on the z axis at points $z = \pm(R/2)$, where R is the internuclear distance in H_2^+ (Fig. 3.1),

$$\hat{H}_e(r\,|\,R) = -\tfrac{1}{2}\nabla_r^2 - \frac{1}{|\mathbf{R}_1 - \mathbf{r}|} - \frac{1}{|\mathbf{R}_2 - \mathbf{r}|} + \frac{1}{R}. \quad (3.102)$$

The energy of an electron occupying the orbital given in Eq. 3.101 can be written as

$$\varepsilon\left(\psi_{H_2^+}\right) = \left\langle \psi_{H_2^+} \middle| \hat{H}_e \middle| \psi_{H_2^+} \right\rangle, \quad (3.103)$$

where

$$\psi_{H_2^+}(r,s)) = \psi(r)\sigma(s) = \left(c_1\varphi_1(r) + c_2\varphi_2(r)\right)\sigma(s). \quad (3.104)$$

[p]Here, for simplicity, we included the term of internuclear repulsion into the electronic Hamiltonian.

The best variational electronic wave function is given by a solution of the equation

$$\mathbf{Hc}_i = \varepsilon_i \mathbf{Sc}_i, \tag{3.105}$$

where $\mathbf{H}$ is the Hamiltonian matrix determined in the basis set of functions $\varphi_1(r)$ and $\varphi_2(r)$

$$\mathbf{H} = \begin{pmatrix} H_{11} & H_{12} \\ H_{21} & H_{22} \end{pmatrix}. \tag{3.106}$$

Here,

$$H_{\mu\nu} = \langle \mu | \hat{H} | \nu \rangle = \int \varphi_\mu^* \hat{H} \varphi_\nu d\mathbf{r}, \tag{3.107}$$

where, in the general case, $H_{\mu\nu} = H_{\nu\mu}^*$, so $\mathbf{H}$ is a self-adjoint matrix. In our case, functions $\varphi_\mu(r)$ are real, so $\mathbf{H}$ is simply a symmetric matrix $H_{\mu\nu} = H_{\nu\mu}$. This is also evident because of the symmetry of the problem (the two hydrogen nuclei are completely equivalent and bear the same AOs). $\mathbf{S}$ is the matrix of overlap integrals such that $S_{11} = S_{22} = 1$ for normalized functions (Eq. 3.101) and

$$S_{12} = S_{21} = \langle \varphi_1 | \varphi_2 \rangle = \int \varphi_1^*(\mathbf{r}) \varphi_2(\mathbf{r}) d\mathbf{r} \tag{3.108}$$

is the overlap integral between functions φ_1 and φ_2. Matrix $\mathbf{S}$ is also symmetric. In Eq. 3.105, ε_i is the ith root of the secular equation and $\mathbf{c}_i$ is the corresponding vector of coefficients $c_{i\mu}$

$$\mathbf{c}_i = \begin{pmatrix} c_{i1} \\ c_{i2} \end{pmatrix}. \tag{3.109}$$

The allowed energy values ε_i, which correspond to the stationary points of the energy functional $E = \langle \varphi | \hat{H} | \varphi \rangle = \int \varphi \hat{H} \varphi d\mathbf{r}$, are given by the solutions of the secular equation

$$\det[\mathbf{H} - \varepsilon \mathbf{S}] = \begin{vmatrix} H_{11} - \varepsilon & H_{12} - \varepsilon S_{12} \\ H_{21} - \varepsilon S_{21} & H_{22} - \varepsilon \end{vmatrix} = 0. \tag{3.110}$$

Equation 3.110 is a quadratic equation in ε with two solutions

$$\varepsilon_1 = \frac{H_{11} + H_{12}}{1 + S_{12}} \tag{3.111}$$

and

$$\varepsilon_2 = \frac{H_{11} - H_{12}}{1 - S_{12}} .$$

Each eigenvalue ε_i is related to the corresponding wave functions $\psi_1(r) = c_{11}\varphi_1(r) + c_{12}\varphi_2(r)$ and $\psi_2(r) = c_{21}\varphi_1(r) + c_{22}\varphi_2(r)$, respectively. Coefficients $c_{i\mu}$ can be found by substituting these roots into the eigenvalue equation (Eq. 3.105) and imposing the following normalization condition:

$$\int |\psi_i(r)|^2 = 1 . \tag{3.112}$$

Finally, we obtain

$$c_{11} = c_{12} = \frac{1}{\sqrt{2 + 2S_{12}}} \tag{3.113}$$

and

$$c_{21} = c_{22} = \frac{1}{\sqrt{2 - 2S_{12}}} .$$

Hence, the wave functions associated with the eigenenergies (Eq. 3.111) can be written as follows:

$$\psi_1(r) = \frac{1}{\sqrt{2 + 2S_{12}}}\left(\varphi_1(r) + \varphi_2(r)\right) \tag{3.114}$$

and

$$\psi_2(r) = \frac{1}{\sqrt{2 - 2S_{12}}}\left(\varphi_1(r) - \varphi_2(r)\right) .$$

The Hamiltonian of H_2^+ (Eq. 3.108) can be represented as a sum of the Hamiltonian of a hydrogen atom plus its interaction with the other proton

$$\hat{H}\left(\mathrm{H}_2^+\right) = \hat{H}_1 + \hat{V}_2 + \hat{V}_{12} = \hat{H}_2 + \hat{V}_1 + \hat{V}_{12} , \tag{3.115}$$

where $\hat{H}_{1(2)}$ is the Hamiltonian of a hydrogen atom located at point 1(2), $\hat{V}_{1(2)}$ designates the (attractive) potential for electrons created by the corresponding proton, and $\hat{V}_{12}$ is the (repulsive) interaction between the two protons. Therefore, taking integrals in Eqs. 3.107 and 3.108, we obtain

$$H_{11} = E_H + \frac{1}{R} - \langle \varphi_1 | V_2 | \varphi_1 \rangle = E_H + \frac{1}{R} - v_{11} , \tag{3.116}$$

where $v_{11} = \int \frac{\varphi_1^2(\mathbf{r})}{|\mathbf{R}_2 - \mathbf{r}|} d\mathbf{r}$ and

$$H_{12} = \left(E_H + \frac{1}{R} \right) S_{12} - \langle \varphi_1 | V_1 | \varphi_2 \rangle = E_H + \frac{1}{R} - v_{12}, \quad (3.117)$$

where $v_{12} = \int \frac{\varphi_1(\mathbf{r})\varphi_2(\mathbf{r})}{|\mathbf{R}_1 - \mathbf{r}|} d\mathbf{r}$. Finally, the eigenvalues ε_1 and ε_2 (Eq. 3.111) will be presented as follows:

$$\varepsilon_1 = E_H + \frac{1}{R} - \frac{v_{11} + v_{12}}{1 + S_{12}} \quad (3.118)$$

and

$$\varepsilon_2 = E_H + \frac{1}{R} - \frac{v_{11} - v_{12}}{1 - S_{12}}.$$

Here, the term $\frac{1}{R}$ describes nuclear repulsion.

Now, we may calculate the integrals v_{11}, v_{12}, and S_{12} directly using the explicit form of the atomic 1s orbital of a hydrogen atom (Eq. 3.101):

$$v_{11} = \left(1 + \frac{1}{R} \right) \exp(-2R), \quad (3.119)$$

$$v_{12} = \left(\frac{1}{R} - \frac{2}{3} R \right) \exp(-R), \quad (3.120)$$

and

$$S_{12} = \left(1 + R + \frac{1}{3} R^2 \right) \exp(-R). \quad (3.121)$$

The overlap integral S_{12} and, in our simple case, the Hamiltonian matrix elements $H_{\mu\nu}$ can be calculated explicitly. Consider that the success of the MO LCAO approximation is determined by the fact that molecular one-electron wave functions are close to atomic ones in the regions of atomic nuclei.

3.3.2 Diatomic Molecules, Their Classification, and Spectroscopy

Diatomic molecules are the simplest molecular systems. Note first that there is now only one internal coordinate in diatomic molecules, namely the interatomic distance $\boldsymbol{R}$, which determines the relative positions of atomic nuclei in the diatomic molecules. Therefore, PESs of diatomic molecules in various electronic states are simply reduced to potential energy curves $E_i(R)$. An example of such a curve for the ground state, $E_0(R)$, is shown in Fig. 3.2.

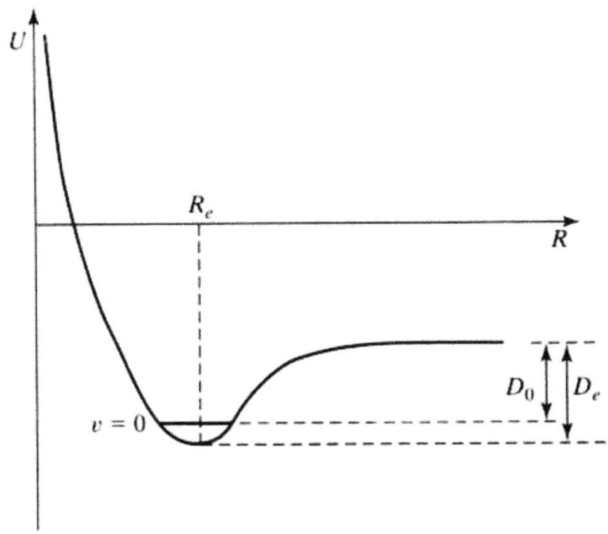

Figure 3.2 Typical potential curve for the ground state of a diatomic molecule.

At infinity (when $R = \infty$), the curve goes to the limit of separated atoms in their ground states. Normally, it has a minimum at an equilibrium distance R_e. A diatomic molecule has only one vibrational mode, whose frequency is determined by the force constant (curvature) at this point. The total ground-state energy of the diatomic molecule is determined by the electronic energy at the equilibrium point $E_0^e = E_0(R_e)$ and ZPE (see Eq. 3.85),

$$E_0^T = E_0^{\mathrm{e}}(R_0) + \tfrac{1}{2}\omega . \tag{3.122}$$

The difference between the energy of separated atoms and the total ground-state energy is the dissociation energy of the diatomic molecule.

Like in the case of atoms, electronic terms of diatomic molecules can be classified by symmetry. The potential energy of a diatomic molecule has only an axial rather than central symmetry. We assume as usual that the nuclear axis of a diatomic molecule is aligned with the z axis. The total orbital angular momentum operator of electrons $\hat{\mathbf{L}}$ does not commute with the Hamiltonian any more. However, the projection of the orbital angular momentum operator onto the molecular axis $\hat{L}_z$ still commutes with the Hamiltonian because of the axial symmetry of the diatomic molecule. Therefore, the eigenfunctions of the electronic Hamiltonian for a diatomic molecule can be classified by the eigenvalues of $\hat{L}_z$, which run over the values $L_z = 0, \pm 1, \pm 2, \ldots$. The absolute values of $\hat{L}_z$ are commonly designated as Λ.

The symmetry of a diatomic molecule include a reflection through any plane containing the nuclear axis z. It is clear that this symmetry operation leaves the total energy unchanged. On the other hand, $\hat{\mathbf{L}}$ changes its sign under this operation, because $\hat{\mathbf{L}}$ is an axial vector. Hence, its projection $\hat{L}_z$ changes its sign as well. This means that such a reflection transforms an electronic eigenfunction characterized by a certain value of L_z to an eigenfunction with the same energy and with the same Λ, but with the opposite sign of L_z. Hence, each electronic term of a diatomic molecule with a nonzero Λ is doubly degenerate. Terms with $\Lambda = 0, 1, 2, \ldots$ are designated by capital Greek letters $\Sigma, \Pi, \Delta, \ldots$, respectively.

Each electronic state is also characterized by its total spin S. For S different from zero, the electronic term is spin-degenerate with a multiplicity of $2S + 1$. The term multiplicity is designated as a left-hand superscript. For example, a term with $\Lambda = 0$ and $S = 0$ will be designated as ${}^1\Sigma$, a term with $\Lambda = 1$ and $S = 1/2$ will be designated as ${}^{1/2}\Pi$, etc.

3.3.3 Classification of MO Shells: Inner (Core), Valence and Pi, and Outer (Virtual) Orbitals

Because in the MO LCAO approximation the MOs are LCAOs, the MOs can be classified according to the AOs that make the main contribution to the particular MO. According to perturbation theory (see Section 3.1.5), the extent to which an orbital is admixed to a zero-order orbital (an eigenfunction of the zero-order Hamiltonian) is determined by the expression $C_{n,\nu}^{(1)} = \frac{\left\langle \Psi_\nu^{(0)} \middle| \hat{V} \middle| \Psi_n^{(0)} \right\rangle}{E_n^{(0)} - E_\nu^{(0)}}$ (Eq. 3.20).

It is evident from this expression that the extent of orbital mixing is determined by two factors: (i) the interaction matrix element $\left\langle \Psi_\nu^{(0)} \middle| \hat{V} \middle| \Psi_n^{(0)} \right\rangle$ (spatial factor) and (ii) the difference in energy between the corresponding levels $E_n^{(0)} - E_\nu^{(0)}$ (energy factor). In their turn, atomic shells of atoms in a molecule can be roughly divided into two groups, the outer shell with the highest n of a given atom and the inner shells, that is, all other shells with lower quantum numbers n.

The orbitals of inner shells are located closer to the particular atomic nucleus, inside the region of the outer shell. Therefore, the inner-shell orbitals that belong to different atoms do not strongly interact with each other because of their spatial separation. Here, we strongly recommend the permanent Internet resource located on the University of Sheffield website [16], on which a gallery of AOs and MOs is presented, including images representing AOs and MOs, animated plots of wave functions, animated plots of electron density, dot-density plots of electron density, and plots of radial distribution functions. On the other hand, AOs that belong to different shells (with different principal quantum numbers n) do not strongly mix in MOs, because, in this case, the energy factor is large.

It follows from this that all MOs in a molecule can be classified as doubly occupied inner-shell (or core) MOs, outer-shell (or valence) MOs, and unoccupied (or virtual) MOs. The valence electrons can further be (approximately) classified into different categories. First, there are so-called bonding MOs, which are responsible for the formation of chemical bonds between atoms, and second, there are so-called nonbonding MOs, or lone-pair orbitals. If a molecule contains planar fragments that are relatively well separated from each other, orbitals in such fragments are commonly classified as sigma (σ) and pi (π) orbitals. Sigma orbitals are localized in the plane of the fragment and describe two-center bonds between pair of atoms. In a local system of coordinates associated with a given fragment such that x and y axes are located in the plane of the planar fragment and the z axis is orthogonal to it, σ orbitals are (mainly) formed from the outer ns, np_x, and np_y orbitals (where n is the principal quantum number of the outer [valence] shell). The π orbitals of a fragment are formed of its outer np_z orbitals, which are directed orthogonal to the fragment along the np_z orbital. The σ orbitals are symmetric with respect to the molecular plane and form the molecular skeleton; the π orbitals are antisymmetric with respect to this plane and may be said to move in the field of the σ electrons and the nuclei.

3.3.4 Construction of Electronic Wave Functions

The MO LCAO approach is based on the independent electron approximation, in which each electron is described by its own one-electron wave function $\psi_i(r_i,s_i) = \phi_i(r_i)\sigma_i(s_i)$, where $\phi_i(r_i)$ is the spatial part of the ith MO (or simply spatial MO) and $\sigma_i(s_i)$ is its spin part. The independence of electrons means that these functions $\psi_i(r_i,s_i)$, and hence, their spatial parts $\phi(r_i)$ are linearly independent.

Spin functions $\sigma(s)$[q] are determined in a 2D space with two basis functions $\alpha(s)$ and $\beta(s)$, which are eigenfunctions of the spin squared operator $\hat{S}^2$ and the spin projection operator $\hat{S}_z$, so $\hat{S}^2|\sigma(s)\rangle = s(s+1)|\sigma(s)\rangle$ and $\hat{S}_z|\sigma(s)\rangle = m_s|\sigma(s)\rangle$, where $s=\frac{1}{2}$ and $m_s = \pm\frac{1}{2}$. The "+" sign corresponds to $|\sigma(s)\rangle = |\alpha(s)\rangle$, while the "–" sign corresponds to $|\sigma(s)\rangle = |\beta(s)\rangle$. In the absence of magnetic fields, it might be considered that each spatial MO can be multiplied by the $\alpha(s)$ or $\beta(s)$ function.

To construct a many-electron wave function from one-electron MOs, it is reasonable to consider a product of spin orbitals $\psi_i(r_i,s_i) = \phi_i(r_i)\sigma_i(s_i)$ (commonly known as the Hartree product [17])

$$\Phi^{HP}(q_i) = \phi_1(r_1)\sigma_1\phi_2(r_2)\,\sigma_2\phi_3(r_3)\,\sigma_3\phi_4(r_4)\,\sigma_4\cdot\ldots\cdot \quad (3.123)$$

[q]We skip here index i for simplicity.

Such a product already possesses many useful properties; however, it does not obey the principle of electron indistinguishability and the Pauli exclusion principle. Both of them can easily be satisfied if the Hartree product (Eq. 3.123) is antisymmetrized by the action of the antisymmetrizer $\hat{A}$

$$\hat{A} = \frac{1}{\sqrt{N!}} \sum_{k}^{N!} (-1)^{p_k} \hat{P}_k , \tag{3.124}$$

where N is the number of electrons, the sum is taken over all permutations $\hat{P}_k$ of a set of N integers 1, . . ., N, and p_k is the parity of permutation $\hat{P}_k$. The total number of terms in Eq. 3.124 is N ! The action of $\hat{A}$ on the simple product (Eq. 3.123) results in a wave function in the form of a Slater determinant (SD)

$$\Phi^{\text{SD}}(q_i) = \hat{A}\Phi^{\text{HP}}(q_i) = \hat{A}\left[\phi_1(r_1)\sigma_1\phi_2(r_2)\sigma_2\phi_3(r_3)\sigma_3\phi_4(r_4)\sigma_4 \cdot \ldots \cdot\right]. \tag{3.125}$$

It might be considered that one-electron MOs are eigenfunctions of some one-electron Hamiltonian (effective Hamiltonian, $\hat{h}^{\text{eff}}$) and form a complete basis set in a certain (sub)space of one-electron functions. Then, all the possible SDs constructed on these MOs form a complete basis set in the space of all antisymmetric N-electron wave functions (Löwdin's expansion theorem [18]).

The distribution of electrons over MOs is called the *electron configuration*. For example, we may consider an electron configuration in which N electrons occupy N lowest spin orbitals (eigenfunctions of $\hat{h}^{\text{eff}}$). Starting from this configuration, one can construct all other configurations by transferring (exciting) one, two, or more electrons from occupied orbitals to virtual (unoccupied) orbitals. In the general case, an electron configuration from which all other configurations are constructed by transferring electrons from occupied to virtual MOs is named the *reference configuration*.

If in a given electron configuration, all spatial orbitals are doubly occupied, and each MO is occupied by two electrons with the opposite spin, this configuration is described by a single SD and is named the *closed-shell configuration*. Otherwise, the electron configuration is called the *open-shell configuration*.

The wave function for a closed-shell configuration with $2N$ electrons can be written as follows:

$$\Phi^{\mathrm{CS}}(r_i\sigma_i)=\hat{A}\begin{bmatrix}\phi_1(r_1)\alpha_1\phi_1(r_2)\beta_2\phi_2(r_3)\alpha_3\phi_2(r_4)\beta_4\cdot\ldots\\ \ldots\cdot\phi_N(r_{2N-1})\alpha_{2N-1}\phi_N(r_{2N})\beta_{2N}\end{bmatrix} \quad (3.126)$$

It is evident that this function is an eigenfunction of operator $\hat{S}_z$ with eigenvalue 0, because it contains an equal number of α and β functions. It is not quite trivial that this function is also an eigenfunction of operator $\hat{S}^2$ with eigenvalue 0 (see, for example, Ref. [19]).

3.4 Roothaan Equations: The SCF MO LCAO Approach

3.4.1 General Methodology: Hartree–Fock Approximation

Consider now an antisymmetrized product of spin orbitals, that is, a single SD

$$\Phi^{\mathrm{SD}}(q_i)=\hat{A}\Phi^{\mathrm{HP}}(q_i)=\hat{A}\left[\phi_1(r_1)\sigma_1\phi_2(r_2)\sigma_2\phi_3(r_3)\sigma_3\phi_4(r_4)\sigma_4\cdot\ldots\cdot\right], \quad (3.127)$$

where the antisymmetrizer operator is determined by Eq. 3.124, and apply to it the variational principle. The idea to replace the Hartree product (Eq. 3.123) by the SD (Eq. 3.127) belongs to V. A. Fock [20], which explains the name of this approximation and the method based on it.

We require $\left\langle\Phi^{\mathrm{SD}}\middle|\hat{H}_{\mathrm{e}}\middle|\Phi^{\mathrm{SD}}\right\rangle=\min$ for all possible spatial orbitals that obey the condition that $\left\langle\phi_i(r)\middle|\phi_j(r)\right\rangle=\delta_{ij}$, where the electronic Hamiltonian $\hat{H}_{\mathrm{e}}(r\,|\,R)=\hat{T}_{\mathrm{e}}(r)+\hat{V}_{\mathrm{ee}}(r)+\hat{V}_{\mathrm{eN}}(r,R)$. Remember here that letters r and R designate the total sets of electronic and nuclear coordinates.

Now, it is necessary to calculate the expression

$$\left\langle\Phi^{\mathrm{SD}}\middle|\hat{H}_{\mathrm{e}}\middle|\Phi^{\mathrm{SD}}\right\rangle=$$

$$\int\hat{A}\left[\phi_1(r_1)\alpha_1\phi_1(r_2)\beta_2\phi_2(r_3)\alpha_3\phi_2(r_4)\beta_4\cdot\ldots\cdot\right]^*\left(\hat{T}_{\mathrm{e}}(r)+\hat{V}_{\mathrm{ee}}(r)+\hat{V}_{\mathrm{eN}}(r,R)\right)$$

$$\hat{A}\left[\phi_1(r_1)\alpha_1\phi_1(r_2)\beta_2\phi_2(r_3)\alpha_3\phi_2(r_4)\beta_4\cdot\ldots\cdot\right]dr_1\ldots dr_N d\sigma_1\ldots d\sigma_N.$$

Using the direct expression for the antisymmetrizer operator (III.3.24), we obtain

$$\left\langle \Phi^{\mathrm{SD}} \middle| \hat{H}_{\mathrm{e}} \middle| \Phi^{\mathrm{SD}} \right\rangle$$

$$= \frac{1}{N!}\int \left(\sum_{k}^{N!} (-1)^{p_k} \hat{P}_k \phi_1(r_1)\alpha_1 \phi_1(r_2)\beta_2 \phi_2(r_3)\alpha_3 \phi_2(r_4)\beta_4 \cdot \ldots \cdot \right)^*$$

$$\times \left(\hat{V}_{\mathrm{NN}}(R) + \hat{T}_{\mathrm{e}}(r) + \hat{V}_{\mathrm{ee}}(r) + \hat{V}_{\mathrm{eN}}(r,R) \right)$$

$$\times \left(\sum_{l}^{N!} (-1)^{p_l} \hat{P}_l \phi_1(r_1)\alpha_1 \phi_1(r_2)\beta_2 \phi_2(r_3)\alpha_3 \phi_2(r_4)\beta_4 \cdot \ldots \cdot \right)$$

$$dr_1 dr_2 \ldots dr_N d\sigma_1 \ldots d\sigma_N.$$

Consider the second sum in this equation. It contains $N!$ terms. Let us act on each term by the inverse permutation $\hat{P}_l^{-1}$, which restore the normal order of variables. This operation does not affect the result of integration, because this result is independent of the order of integration. Hence, we will have $N!$ identical terms, and we may write

$$\left\langle \Phi^{\mathrm{SD}} \middle| \hat{H}_{\mathrm{e}} \middle| \Phi^{\mathrm{SD}} \right\rangle$$

$$= \int \left(\sum_{k}^{N!} (-1)^{p_k} \hat{P}_k \phi_1(r_1)\alpha_1 \phi_1(r_2)\beta_2 \phi_2(r_3)\alpha_3 \phi_2(r_4)\beta_4 \cdot \ldots \cdot \right)^*$$

$$\times \left(\hat{T}_{\mathrm{e}}(r) + \hat{V}_{\mathrm{ee}}(r) + \hat{V}_{\mathrm{eN}}(r,R) \right)$$

$$\times \left(\phi_1(r_1)\alpha_1 \phi_1(r_2)\beta_2 \phi_2(r_3)\alpha_3 \phi_2(r_4)\beta_4 \cdot \ldots \cdot \right)$$

$$dr_1 dr_2 \ldots dr_N d\sigma_1 \ldots d\sigma_N.$$

It is convenient to present the electronic Hamiltonian as a sum of its one-electron and two-electron parts, $\hat{H}_{\mathrm{e}} = \hat{h}^{(1)} + \hat{h}^{(2)}$, where $\hat{h}^{(1)} = \hat{T}_{\mathrm{e}}(r) + \hat{V}_{\mathrm{eN}}(r,R)$, and $\hat{h}^{(2)} = \hat{V}_{\mathrm{ee}}(r) = \sum_{m \neq n} r_{mn}^{-1}$. Then, it is rather easy to show that $\left\langle \Phi^{\mathrm{SD}} \middle| \hat{h}^{(1)} \middle| \Phi^{\mathrm{SD}} \right\rangle = \sum_{i=1}^{N} \left\langle \psi_i \middle| \hat{h}^{(1)} \middle| \psi_i \right\rangle = \sum_{i=1}^{N} \hat{h}_{ii}^{(1)}$. Similarly, it may be shown that $\left\langle \Phi^{\mathrm{SD}} \middle| \hat{h}^{(2)} \middle| \Phi^{\mathrm{SD}} \right\rangle = \frac{1}{2} \sum_{i,j}^{N} \left(\left\langle ij \middle| r_{12}^{-1} \middle| ij \right\rangle - \left\langle ji \middle| r_{12}^{-1} \middle| ij \right\rangle \right)$. Finally, we obtain

$$\left\langle \Phi^{\text{SD}} \middle| \hat{H}_{\text{e}} \middle| \Phi^{\text{SD}} \right\rangle = \sum_{i=1}^{N} \hat{h}_{ii}^{(1)} + \tfrac{1}{2} \sum_{i,j}^{N} \left(\langle ij | r_{12}^{-1} | ij \rangle - \langle ji | r_{12}^{-1} | ij \rangle \right). \quad (3.128)$$

Here, we introduced the designation $\langle ij | r_{12}^{-1} | kl \rangle = \int \psi_i^*(2) \psi_j^*(1) r_{12}^{-1} \psi_k(2) \psi_l(1) dr_1 dr_2$.

To find the spatial orbitals that correspond to a minimum of functional $\left\langle \Phi^{\text{SD}} \middle| \hat{H}_{\text{e}} \middle| \Phi^{\text{SD}} \right\rangle$ with the restriction that $\langle \phi_i(r) | \phi_j(r) \rangle = \delta_{ij}$, it is common to use the method of Lagrange undetermined multipliers and to minimize the Lagrange functional

$$L\left[\{\psi_i\}\right] = \left\langle \Phi^{\text{SD}}\{\psi_i\} \middle| \hat{H}_{\text{e}} \middle| \Phi^{\text{SD}}\{\psi_i\} \right\rangle - \sum_{ij} \varepsilon_{ij} \left(\langle \psi_i | \psi_j \rangle - \delta_{ij} \right). \quad (3.129)$$

This approach is named the Hartree–Fock (HF) method, and the corresponding optimal single-determinant wave function is named the HF wave function. Now, we should calculate the variational derivative of functional $L[\{\psi_i\}]$ and assume that it equals zero:

$$\begin{aligned} &\frac{\delta L\left[\{\psi_i\}\right]}{\delta \psi_i^*} \\ &= \frac{\delta}{\delta \psi_i^*} \left(\sum_{i=1}^{N} \hat{h}_{ii}^{(1)} + \tfrac{1}{2} \sum_{i,j}^{N} \left(\langle ij | r_{12}^{-1} | ij \rangle - \langle ji | r_{12}^{-1} | ij \rangle \right) \right) \\ &- \frac{\delta}{\delta \psi_i^*} \sum_{ij} \varepsilon_{ij} \langle \psi_i | \psi_j \rangle = 0 \end{aligned}$$

The variational derivative of the first term in parentheses equals

$$\frac{\delta}{\delta \psi_i^*} \sum_{i=1}^{N} \hat{h}_{ii}^{(1)} = \frac{\delta}{\delta \psi_i^*} \sum_{i=1}^{N} \langle \psi_i | \hat{h}^{(1)} | \psi_i \rangle = \hat{h}^{(1)} | \psi_i \rangle,$$

and the variational derivative of the second term in parentheses equals

$$\begin{aligned} &\frac{\delta}{\delta \psi_i^*} \left(\tfrac{1}{2} \sum_{i,j}^{N} \left(\langle ij | r_{12}^{-1} | ij \rangle - \langle ji | r_{12}^{-1} | ij \rangle \right) \right) \\ &= \frac{\delta}{\delta \psi_i^*} \left(\tfrac{1}{2} \sum_{i,j}^{N} \left(\begin{aligned} &\int \psi_i^*(2) \psi_j^*(1) r_{12}^{-1} \psi_i(2) \psi_j(1) dr_1 dr_2 \\ &- \int \psi_j^*(2) \psi_i^*(1) r_{12}^{-1} \psi_i(2) \psi_j(1) dr_1 dr_2 \end{aligned} \right) \right) \end{aligned}$$

$$= \left(\sum_j^N \left(\begin{array}{l} \int \psi_j^*(1) r_{12}^{-1} \psi_j(1) dr_1 \psi_i(2) \\ - \int \psi_j^*(2) r_{12}^{-1} \psi_i(2) \psi_j(1) dr_2 \end{array} \right) \right) = \left(\hat{J} - \hat{K} \right) |\psi_i\rangle .$$

Here $\hat{J}$ is the Coulomb multiplication operator by function $J(r_2) = \sum_j^N \int \psi_j^*(r_1) r_{12}^{-1} \psi_j(r_1) dr_1$ and $\hat{K}$ is the integral exchange operator such that $\hat{K}\psi_i = \int K(r_1, r_2) \psi_i(r_2) dr_2$, where the kernel $K(r_1, r_2) \quad \psi_j^*(2) r_{12}^{1} \psi_j(1)$.

Finally, we obtain the following HF equations in the form

$$\hat{h}^{(1)} + \hat{J} - \hat{K} |\psi_i\rangle = \sum_j \boldsymbol{\epsilon}_{ij} |\psi_j\rangle . \tag{3.130}$$

It might be easily shown that the matrix $\boldsymbol{\varepsilon}_{ij}$ is Hermitian and can be diagonalized by an unitary transformation of vectors $|\boldsymbol{\psi}_j\rangle$. Hence, we obtain the HF equations in the form of the eigenvalue problem

$$\hat{h}^{(1)} + \hat{J} - \hat{K} |\psi_i\rangle = \varepsilon_i |\psi_i\rangle . \tag{3.131}$$

The one-electron operator $\hat{F} = \hat{h}^{(1)} + \hat{J} - \hat{K}$ is called the HF operator or simply *Fockian*.

The specificity of this problem is that the operator in the left part of this equation itself depends on the solution ψ_i, that is, equations and solutions should be self-consistent. In other words, the field in which electrons move should be consistent with their orbitals. Therefore, the HF method is also named the self-consistent field molecular orbital (SCF MO) method.

HF equations are integro-differential ones and cannot be solved for molecular systems even numerically. A computationally implementable approach consists of the use of the LCAO approximation for the presentation of the SCF MOs and to consider the coefficients of the LCAO expansion as a finite set of variational parameters. In this case, the integro-differential HF equations are reduced to a system of nonlinear algebraic equations. Such an approach was first proposed by C. C. J. Roothaan for the case of closed shells [21] and further extended to the case of open shells

[22]. Therefore, it is commonly named the Hartree–Fock–Roothaan (HFR) method. In the next sections, we will consider separately both cases.

3.4.2 LCAO Approximation for Molecular Systems

As well as in the simplest case of the hydrogen molecular ion, we will consider that MOs of a general molecular system can be presented as a linear combination of its AOs

$$\phi_i = \sum_{\mu} \chi_\mu C_{\mu i}\,, \tag{3.132}$$

where χ_μ are normalized AOs, which form the basis set in which all MOs are expanded.

For a spherically symmetric atom, AOs can be presented as $\chi_\mu = R_{nl}(r)Y_{lm}(\theta,\varphi)$. Here, $R_{nl}(r)$ is the radial function $R_{nl}(r)=\sqrt{\left(\frac{2Z}{na_\mu}\right)^3 \frac{(n-l-1)!}{2n[(n+l)!]}} e^{-Zr/na_\mu}\left(\frac{2Zr}{na_\mu}\right)^l L_{n-l-1}^{2l+1}\left(\frac{2Zr}{na_\mu}\right)$, where L_{n-l-1}^{2l+1} are generalized Laguerre polynomials of degree $(n - l - 1)$, $a_\mu = \frac{4\pi\varepsilon_0\hbar^2}{\mu e^2} = \frac{m_e}{\mu}a_0$, and μ is the reduced mass of the electron-nucleus system $\mu \approx m_e$. That is, $R_{nl}(r) \propto r^{n-1}e^{-Zr/na_\mu}$, and $Y_{lm}(\theta,\varphi)$ is a common spherical harmonic. Historically, the radial parts of AOs in LCAO theory were selected as $R_{nl}(r) = Ar^{n-1}e^{-Zr/na_\mu}$ (Slater-type orbitals, STOs). STOs can also be used in the Cartesian form $\chi_{\text{STO}}(\alpha) = Nx^a y^b z^c \exp(-\alpha r)$. The important advantage of STOs is that these orbitals obey both asymptotic conditions at $r \to \infty$ ($\psi_i \sim \exp\left[-\left(\sqrt{\varepsilon_i}\right)r\right]$, where ε_i is the orbital energy) and at $r \to 0$ (the so-called cusp condition, which states that the electron density $\rho(r)$ has a cusp at the position of the nucleus $\left.\frac{d\rho(r)}{dr}\right|_{r\to 0} = -\frac{2\rho(r)Z_k}{a_0}$). However, it was rather quickly found that the use of STOs leads to great computational problems in molecular calculations. These difficulties can be avoided with the use of Gaussian-type orbitals (GTOs) proposed by

S. F. Boys [23]. GTOs are commonly used in the Cartesian form $\chi_{\text{GTO}}(\alpha) = Nx^a y^b z^c \exp(-\alpha r^2)$.

A useful property of GTOs is the so-called Gaussian product theorem, stating that a product of two GTOs is again a new GTO whose center is on the line connecting the centers of initial GTOs. It might be easily proven for the case of Gaussian s orbitals, that is, for the case $a = b = c = 0$. Actually, let $\chi^1_{\text{GTO}}(\alpha_1) = N^1 \exp\left(-\alpha_1 (r - R_1)^2\right)$ and $\chi^2_{\text{GTO}}(\alpha_2) = N^2 \exp\left(-\alpha_2 (r - R_2)^2\right)$.

Then

$$\chi^1_{\text{GTO}}(\alpha_1)\chi^2_{\text{GTO}}(\alpha_2) = N^1 N^2 \exp\left(-\alpha_1 (r - R_1)^2 - \alpha_2 (r - R_2)^2\right)$$
$$= N^1 N^2 \exp\left(-(\alpha_1 + \alpha_2)(r - R_3)^2\right)\exp\left(-\alpha_3 (R_1 - R_2)^2\right).$$

Here, $R_3 = \dfrac{\alpha_1 R_1 + \alpha_2 R_2}{\alpha_1 + \alpha_2}$, and $\alpha_3 = \dfrac{\alpha_1 \alpha_2}{\alpha_1 + \alpha_2}$.

3.4.3 Closed Shells: Hartree–Fock–Roothaan Equations

Let us start from the MO LCAO expansion

$$\phi_i = \sum_{\mu} \chi_\mu C_{\mu i}, \tag{3.133}$$

where normalized AOs χ_μ, $\langle \chi_\mu | \chi_\mu \rangle = \int \chi^*_\mu \chi_\mu dv = 1$, are centered on atoms A_k. We introduce the following matrix designations: $\boldsymbol{\chi}$ is the row vector (χ_μ), $\mathbf{c}_i$ are column vectors $\{C_{\mu i}\}$, and $\mathbf{C}$ is the matrix of coefficients $\{C_{\mu i}\}$. Then, $\phi_i = \boldsymbol{\chi}\mathbf{c}_i$, $\mathbf{f} = \boldsymbol{\chi}\mathbf{C}$. Let $\hat{M}$ be a one-electron operator. We can determine its matrix elements in terms of basis set functions χ_p, $M_{pq} = \int \chi^*_p M \chi_q dv$, or $\mathbf{M} = [M_{pq}]$. Correspondingly, we define matrices $\mathbf{h}$, $\mathbf{J}_i$, $\mathbf{k}_i$, $\mathbf{G}$, and $\mathbf{F}$ for operators $\hat{h}$, $\hat{J}_i$, $\hat{K}_i$, $\hat{G}$, and $\hat{F}$. We also define the matrix of overlap integrals (or simply, the overlap matrix) $\mathbf{S} = [S_{\mu\nu}]$, where $S_{\mu\nu} = \int \chi^*_\mu \chi_\nu dv$. In terms of matrix operations, matrix elements of operator $\hat{M}$ in the MO basis can be presented as

$$\int \chi^*_\mu M \chi_\nu = \overline{\mathbf{c}}_i \mathbf{M} \mathbf{c}_j. \tag{3.134}$$

Here and from here on, the bar over a letter means Hermitian conjugation, while the star means complex conjugation. Then, the condition of MO orthonormality can be written as

$$\int \phi_i^* \phi_j dv = \overline{\mathbf{c}}_i \mathbf{S} \mathbf{c}_j = \delta_{ij} \tag{3.135}$$

Similarly, for the matrix elements of individual terms of the HFR operator, we obtain

$$h_i = \overline{\mathbf{c}}_i \mathbf{h} \mathbf{c}_i \tag{3.136}$$

$$J_{ij} = \overline{\mathbf{c}}_i \mathbf{J}_j \mathbf{c}_i = \overline{\mathbf{c}}_j \mathbf{J}_i \mathbf{c}_j \tag{3.137}$$

$$K_{ij} = \overline{\mathbf{c}}_i \mathbf{K}_j \mathbf{c}_i = \overline{\mathbf{c}}_j \mathbf{K}_i \mathbf{c}_j \tag{3.138}$$

Let us calculate the variation of the expression for the total energy of the HFR wave function with respect to coefficients $C_{\mu i}$.

$$\begin{aligned}\delta E &= 2\sum_i \delta h_i + \sum_{ij}\left(2\delta J_{ij} - \delta K_{ij}\right)\\ &= 2\sum_i (\delta\overline{\mathbf{c}}_i)\mathbf{h}\mathbf{c}_i + \sum_{ij}\left\{(\delta\overline{\mathbf{c}}_i)(2\mathbf{J}_j - \mathbf{K}_j)\mathbf{c}_i + (\delta\overline{\mathbf{c}}_j)(2\mathbf{J}_i - \mathbf{K}_i)\mathbf{c}_j\right\}\\ &+2\sum_i \overline{\mathbf{c}}_i\mathbf{h}(\delta\mathbf{c}_i) + \sum_{ij}\left\{\overline{\mathbf{c}}_i(2\mathbf{J}_j - \mathbf{K}_j)(\delta\mathbf{c}_i) + \overline{\mathbf{c}}_j(2\mathbf{J}_i - \mathbf{K}_i)(\delta\mathbf{c}_j)\right\}\end{aligned} \tag{3.139}$$

Let us write the equation for the variation δE in a more compact form

$$\begin{aligned}\delta E &= 2\sum_i (\delta\overline{\mathbf{c}}_i)\left\{\mathbf{h} + \sum_j (2\mathbf{J}_j - \mathbf{K}_j)\right\}\mathbf{c}_i\\ &+2\sum_i (\delta\overline{\mathbf{c}}_i)\left\{\overline{\mathbf{h}} + \sum_j (2\overline{\mathbf{J}}_j - \overline{\mathbf{K}}_j)\right\}\overline{\mathbf{c}}_i\end{aligned} \tag{3.140}$$

and introduce the designation

$$\mathbf{F} = \mathbf{h} + \sum_j (2\mathbf{J}_j - \mathbf{K}_j). \tag{3.141}$$

Then, the expression for δE becomes simply $\delta E = 2\sum_i (\delta\mathbf{c}_i)\mathbf{F}\mathbf{c}_i$ $+2\sum_i (\delta\overline{\mathbf{c}}_i)\overline{\mathbf{F}}\overline{\mathbf{c}}_i$. From the orthonormality condition, we can obtain the condition for $\delta \mathbf{c}_i$

$$(\delta \bar{\mathbf{c}}_i)\mathbf{S}\mathbf{c}_j + \bar{\mathbf{c}}_i\mathbf{S}(\delta \mathbf{c}_j) = 0. \tag{3.142}$$

Using again the method of Lagrange undetermined multipliers, we construct the Lagrange functional by multiplying the orthonormality condition (Eq. 3.142) by Lagrangian multipliers $-2\varepsilon_{ij}$ and summing the obtained equations over indices i and j. This gives $-2\sum_{ij}(\delta \bar{\mathbf{c}}_i)\mathbf{S}\mathbf{c}_j\varepsilon_{ij} - 2\sum_{ij}\bar{\mathbf{c}}_i\mathbf{S}(\delta \mathbf{c}_j)\varepsilon_{ij} = 0$ or

$$-2\sum_{ij}(\delta \mathbf{c}_i^*)\mathbf{S}\mathbf{c}_j\varepsilon_{ji} - 2\sum_{ij}(\delta \mathbf{c}_i^T)\bar{\mathbf{S}}\bar{\mathbf{c}}_j\varepsilon_{ij} = 0. \tag{3.143}$$

Adding expression in the left-hand side of Eq. 3.143 to the expression for the variation of energy, we obtain

$$\delta E' = 2\sum_i(\delta \mathbf{c}_i^*)\left(\mathbf{F}\mathbf{c}_i - \sum_j \mathbf{S}\mathbf{c}_j\varepsilon_{ji}\right) + 2\sum_i(\delta \mathbf{c}_i^T)\left(\bar{\mathbf{F}}\bar{\mathbf{c}}_i - \sum_j \bar{\mathbf{S}}\bar{\mathbf{c}}_j\varepsilon_{ij}\right).$$

The condition that $\delta E' = 0$ for an arbitrary choice of vectors δc_i and $\delta \bar{\mathbf{c}}_i$ or $\delta \mathbf{c}_i^T$ and $\delta \mathbf{c}_i^*$ leads to equations

$$\left.\begin{aligned} \mathbf{F}\mathbf{c}_i &= \sum_j \mathbf{S}\mathbf{c}_j\varepsilon_{ji} \\ \bar{\mathbf{F}}\bar{\mathbf{c}}_i &= \sum_j \bar{\mathbf{S}}\bar{\mathbf{c}}_j\varepsilon_{ij} \end{aligned}\right\}. \tag{3.144}$$

It might be shown that ε_{ij} are elements of an Hermitian matrix, so the obtained equations in Eq. 3.145 are equivalent. The Hermitian matrix ε can be diagonalized by a unitary transformation, and we obtain the final HFR equation

$$\mathbf{F}\mathbf{c}_i = \varepsilon_i\,\mathbf{S}\mathbf{c}_i \text{ or } (\mathbf{F} - \varepsilon\mathbf{S})\mathbf{c} = 0.$$

The allowed values of ε are determined from the secular equation $|\mathbf{F} - \varepsilon\mathbf{S}| = 0$.

3.4.4 Closed Shells: Ionization Potentials and Koopmans's Theorem

Let us consider a singly ionized state obtained upon removing one of two electrons occupying orbital ϕ_i. The corresponding wave function might be written as

$$
{}^2\Phi_i = (\phi_1\alpha)(\phi_1\beta)\cdots(\phi_{i-1}\alpha)(\phi_{i-1}\beta)
\times\begin{Bmatrix}(\phi_i\alpha)\\(\phi_i\beta)\end{Bmatrix}(\phi_{i+1}\alpha)(\phi_{i+1}\beta)\cdots(\phi_n\alpha)(\phi_n\beta),
$$

and the corresponding energy would be given by the expression

$$
\begin{aligned}
E({}^2\Phi_i) &= 2\sum_{j\neq i} H_j + H_i + \sum_{j,k\neq i}(2J_{jk}-K_{jk}) + \sum_{j\neq i}(2J_{ij}-K_{ij})\\
&= 2\sum_j H_j + \sum_{jk}(2J_{jk}-K_{jk}) - H_i - \sum_i (2J_{ij}-K_{ij})\\
&= E({}^1\Phi_0) - H_i - \sum_i (2J_{ij}-K_{ij}).
\end{aligned}
$$

Hence, the energy required for removing one electron from orbitals ϕ_i is given by the equation

$$
E({}^2\Phi_i) - E({}^1\Phi_0) = -H_i - \sum_i (2J_{ij}-K_{ij}).
$$

This expression can be presented in a more convenient form both in the HF method and in the LCAO version. In the case of the HF method, we obtain

$$
H_i + \sum_i (2J_{ij}-K_{ij}) = \int \bar{\phi}_i \left\{ H + \sum_j (2J_j - K_j) \right\} \phi_i = \int \bar{\phi}_i F \varphi_i = \varepsilon_i .
$$

In the case of the LCAO method, we also obtain

$$
H_i + \sum_i (2J_{ij}-K_{ij}) = \mathbf{c}_i^* \left\{ \mathbf{H} + \sum_j (2\mathbf{J}_j - \mathbf{K}_j) \right\} \mathbf{c}_i = \mathbf{c}_i^* \mathbf{F} \mathbf{c}_i = \varepsilon_i \mathbf{c}_i^* \mathbf{S} \mathbf{c}_i = \varepsilon_i .
$$

That is, the energy of electron detachment from orbital *i* equals its energy taken with an opposite sign. Thus, HF theory, both in the HF and the LCAO method, we obtain that the orbital ionization potential (IP) equals the corresponding orbital energy taken with an opposite sign:

$$
E({}^2\Phi_i) - E({}^1\Phi_0) = -\varepsilon_i . \tag{3.145}
$$

This result was first obtained by Dutch American mathematician Tjalling Koopmans[r] [24]. Therefore, the corresponding statement is called Koopmans's theorem.

[r] It is interesting that T. Koopmans won the Nobel Prize for economics in 1975 jointly with Soviet scientist L. Kantorovich.

A similar result can also be obtained for states described by singly excited wave functions, if it is assumed that orbitals remain unchanged upon excitation. Consider excitation $\phi_i \rightarrow \phi_a$. Now, orbitals ϕ_i and ϕ_a will be occupied by one electron each. These two spatial orbitals give rise to four spin orbitals, on which two electrons can be arranged by four ways under the condition that each spatial orbital is singly occupied. Symbolically, this can be written as follows:

$$(\varphi_1\alpha)(\varphi_1\beta)\cdots(\varphi_{i-1}\alpha)(\varphi_{i-1}\beta)\begin{Bmatrix} \varphi_i\alpha & \varphi_a\alpha \\ \varphi_i\alpha & \varphi_a\beta \\ \varphi_i\beta & \varphi_a\alpha \\ \varphi_i\beta & \varphi_a\beta \end{Bmatrix}.$$
$$\times(\varphi_{i+1}\alpha)(\varphi_{i+1}\beta)\cdots(\varphi_n\alpha)(\varphi_n\beta)$$

Four determinant functions give rise to one singlet function with spin $S = 0$ and three degenerate components of a triplet state with spin $S = 1$

$$\left.\begin{array}{l} {}^1\Phi_{i\alpha} = (\varphi_1\alpha)\cdots(\varphi_{i-1}\beta) \\ \times\dfrac{1}{\sqrt{2}}\{(\varphi_i\alpha)(\varphi_a\beta)-(\varphi_i\beta)(\varphi_a\alpha)\}\times(\varphi_{i+1}\alpha)\cdots(\varphi_n\beta) \\ {}^3\Phi_{i\alpha} = (\varphi_1\alpha)\cdots(\varphi_{i-1}\beta) \\ \times\begin{Bmatrix} (\varphi_i\alpha)(\varphi_a\alpha) \\ \dfrac{1}{\sqrt{2}}\{(\varphi_i\alpha)(\varphi_a\beta)+(\varphi_i\beta)(\varphi_a\alpha)\} \\ (\varphi_i\beta)(\varphi_a\beta) \end{Bmatrix}\times(\varphi_{i+1}\alpha)\cdots(\varphi_n\beta) \end{array}\right\}$$

For the corresponding excitation energies, the following simple result can be obtained:

$$E\left({}^{1,3}\Phi_{ia}\right)-E\left({}^1\Phi_0\right)$$
$$=H_a+\sum_j\left(2J_{ja}-K_{ja}\right)-H_i-\sum_j\left(2J_{ij}-K_{ij}\right)-\left(J_{ja}-K_{ja}\right)\pm K_{ia}$$

Here the "+" sign corresponds to the singlet state, while the "–" sign corresponds to the triplet state. Taking into account the equation for orbital energies, we finally obtain

$$E({}^{1,3}\Phi_{ia}) - E({}^1\Phi_0) = \varepsilon_a - \varepsilon_i - (J_{ia} - K_{ia}) \pm K_{ia}$$

3.4.5 Open Shells: Restricted and Unrestricted Treatment

The term "open shell" relates to electron configurations that contain singly occupied spatial orbitals, that is, some number of unpaired electrons. Typical examples of corresponding molecular systems are radicals (systems with one unpaired electron, $S_z = \pm\frac{1}{2}$, $S = \frac{1}{2}$) and triplet molecules ($S_z = \pm 1, S = 1$). In open-shell systems, the number of electrons with spin α (N_α) differs from the number of electrons with spin β (N_β). For definiteness, assume that $N_\alpha > N_\beta$.

There are two main possibilities for construction of wave functions for open-shell systems. In the first one, the set of MOs contains a set of doubly occupied closed-shell MOs $\{\phi_c\}$ and a set of partially occupied open-shell MOs $\{\phi_o\}$. The total set of MOs is $\phi = (\phi_c, \phi_o)$, and orbitals ϕ_i are assumed to be orthonormal and mutually orthogonal. As distinct from the closed shall case, the total electronic wave function is, in general, not described by a single SD, but represents a linear combination of several SDs. Each SD in this linear combination is constructed from a set of doubly occupied closed-shell orbitals, common for all SDs in this expansion, and partially occupied open-shell orbitals selected from $\{\phi_o\}$. Such an approach is called *restricted open-shell theory*. The wave function in the restricted open-shell formalism is an eigenfunctions of $\hat{S}_z$, total spin $\hat{S}$, and $\hat{S}^2$ operators. In the second approach, it is assumed that electrons with spins α and β are, in principle, described by different spatial orbitals. This approach is called *unrestricted open-shell theory*. The SD constructed from such MOs in the unrestricted open-shell formalism is still an eigenfunctions of $\hat{S}_z$ but is not now an eigenfunctions of total spin $\hat{S}$ and $\hat{S}^2$ operators.

3.4.6 Restricted Open-Shell Hartree–Fock–Roothaan Equations

The HF equations for the restricted open-shell formalism were first developed in the fundamental paper by C. C. J. Roothaan [23] and further developed in two important papers by S. Huzinaga [25, 26]. It should be specially noted that the presentation of open-shell theory given in Ref. [23] still remains one of the most clear and exhaustive presentations as compared to more recent works (for a development of high-symmetry molecules, see Ref. [27]). We briefly present here only a general scheme of the original formalism given in Ref. [23]. The method is called *restricted open-shell Hartree–Fock method*, commonly abbreviated as ROHF.

The expectation value of the total electronic energy is given by the following general expression:

$$E = 2\sum_{k} H_k + \sum_{kl}(2J_{kl} - K_{kl})$$
$$+ f\left[2\sum_{m} H_m + f\sum_{mn}(2aJ_{mn} - bK_{mn})\right] + 2\sum_{km}(2J_{km} - K_{km}), \quad (3.146)$$

where indices k, l are used for the closed-shell orbitals; m, n for open-shell orbitals; and i, j for orbitals of an unspecified set. Parameters a, b, and f depend on the particular case. The number f is the fraction of occupied open-shell spin orbitals, that is, $0 < f < 1$. Integrals H_k, J_{kl}, and K_{kl} have the same meaning as in closed-shell theory: these are one-electron matrix elements, and two-electron Coulomb and exchange integrals. Then, Coulomb and exchange operators associated, respectively, with the closed and open shells are introduced as

$$\hat{J}_{\mathrm{C}} = \sum_{k} \hat{J}_k, \; \hat{J}_{\mathrm{O}} = f\sum_{m} \hat{J}_m, \; \hat{J}_{\mathrm{T}} = \hat{J}_{\mathrm{C}} + \hat{J}_{\mathrm{O}}$$

and

$$\hat{K}_{\mathrm{C}} = \sum_{k} \hat{K}_k, \; \hat{K}_{\mathrm{O}} = f\sum_{m} \hat{K}_m, \; \hat{K}_{\mathrm{T}} = \hat{K}_{\mathrm{C}} + \hat{K}_{\mathrm{O}}. \quad (3.146)$$

Now, the following Hermitian Coulomb and exchange coupling operators associated with a certain orbital ϕ_i are defined:

$$\hat{L}_i\phi = \langle\phi_i|\hat{J}_{\mathrm{O}}|\phi\rangle\phi_i + \langle\phi_i|\phi\rangle\hat{J}_{\mathrm{O}}\phi_i$$
$$\hat{M}_i\phi = \langle\phi_i|\hat{K}_{\mathrm{O}}|\phi\rangle\phi_i + \langle\phi_i|\phi\rangle\hat{K}_{\mathrm{O}}\phi_i \quad (3.147)$$

The corresponding closed-shell, open-shell, and total Coulomb and exchange coupling operators are defined as well:

$$\hat{L}_{\mathrm{C}} = \sum_{k} \hat{L}_k, \; \hat{L}_{\mathrm{O}} = f\sum_{m} \hat{L}_m, \; \hat{L}_{\mathrm{T}} = \hat{L}_{\mathrm{C}} + \hat{L}_{\mathrm{O}}$$
$$\hat{M}_{\mathrm{C}} = \sum_{k} \hat{M}_k, \; \hat{M}_{\mathrm{O}} = f\sum_{m} \hat{M}_m, \; \hat{M}_{\mathrm{T}} = \hat{M}_{\mathrm{C}} + \hat{M}_{\mathrm{O}} \quad (3.148)$$

Operators defined in Eqs. 3.146 and 3.148 are invariant under a unitary transformation of both closed-shell and open-shell MOs among themselves separately.

Then, it is necessary to minimize Eq. 3.145 with the restriction that orbitals ϕ_i of the total set of MOs (ϕ_C, ϕ_O) are assumed to be orthonormal and mutually orthogonal. The use of the usual technique of Lagrangian multipliers results in a system of coupled equations for closed- and open-shell orbitals

$$\hat{F}_C \phi_C = \phi_C \eta_C \tag{3.149a}$$

and

$$\hat{F}_O \phi_O = \phi_O \eta_O, \tag{3.149b}$$

where Fockians for the closed- and open-shell orbitals are expressed as follows:

$$F_C = H + 2J_C - K_C + 2J_O - K_O + 2\alpha L_O - \beta M_O \tag{3.150a}$$

and

$$F_O = H + 2J_C - K_C + 2aJ_O - bK_O + 2\alpha L_C - \beta M_C. \tag{3.150b}$$

Matrix elements of Hermitian matrices $\boldsymbol{\eta}_C$ and $\boldsymbol{\eta}_O$ are Lagrangian multipliers θ_{lk} and θ_{nm}/f. Matrices $\boldsymbol{\eta}_C$ and $\boldsymbol{\eta}_O$ can be diagonalized using unitary transformations of closed-shell orbitals and open-shell orbitals. We arrive at orbital equations

$$\hat{F}_C \phi_k = \eta_k \phi_k \tag{3.151a}$$

and

$$\hat{F}_O \phi_m = \eta_m \phi_m. \tag{3.151b}$$

As well as in the case of closed shells, the total energy of this open-shell system can be expressed in terms of orbital energies η_i and one-electron integrals H_i as follows:

$$E = \sum_k (H_k + \eta_k) + f \sum_m (H_m + \eta_m) \tag{3.152}$$

As distinct from the case of closed shells, the eigenenergies η_k and η_m are not generally equal to orbital IPs taken with the opposite sign. This means that there is no simple analogue of Koopmans's theorem for open-shell systems. For more details, see Refs. [28–31].

3.5 Orbital Methods for Treatment of Excited States and Electron Correlation in Molecules

3.5.1 Configuration and Multiconfiguration Interaction: CIS, CISD, and CAS

The main disadvantage of the HF method and its LCAO version, the HFR method, is the disregard of so-called electron correlation, that is, the dependence of the probability to find an electron in a given point on the particular arrangement of all other electrons. This disadvantage is a consequence of the presentation of the many-electron wave function in the form of a single SD or as a linear combination (as in the ROHF method) of a small number of SDs. The natural next step is to consider a space of all SDs that can be constructed from all available spin orbitals. Because of the linear property of determinants, which is well known from linear algebra, this set forms a linear normalized space.

Let us now use the variational principle (see Section 3.1.1) in order to find the best approximate electron wave function in this space. We start from a reference wave function (see Section 3.3.4), which might be, but not obligatory, selected as an HF determinant. Then we consider all possible determinants that can be constructed by transferring electrons (one or more) from an occupied MO to a virtual MO and will search the wave function $|\Psi\rangle$ in the form of a linear combination of the ground (reference) configuration $|\Phi_0\rangle$ and all excited configurations (see, for example, Ref. [29]).

$$|\Psi\rangle = c_0|\Phi_0\rangle + \sum_{ia} c_i^a|\Phi_i^a\rangle + \sum_{a<b,i<j} c_{ij}^{ab}|\Phi_{ij}^{ab}\rangle + \sum_{a<b<c,i<j<k} c_{ijk}^{abc}|\Phi_{ijk}^{abc}\rangle + \ldots \quad (3.153)$$

Here $|\Phi_i^a\rangle$ is the SD that is formed from the reference SD $|\Phi_0\rangle$ by replacing spin orbitals i occupied in $|\Phi_0\rangle$ with a virtual spin orbital a. The designations like $|\Phi_{ij}^{ab}\rangle$, $|\Phi_{ijk}^{abc}\rangle$, and so on are self-evident. Such an approach is called *configuration interaction (CI)*, and functions $|\Phi\rangle$ in Eq. 3.153 are called *configuration functions.* We may briefly write Eq. 3.153 as

$$\left|\Psi^{\mathrm{CI}}\right\rangle = \sum_i c_i \left|\Phi_i\right\rangle. \qquad (3.154)$$

The variational problem is formulated as

$$\left\langle\Psi^{\mathrm{CI}}\middle|\hat{H}_{\mathrm{e}}\middle|\Psi^{\mathrm{CI}}\right\rangle = \min, \qquad (3.155)$$

with the restriction that function Ψ^{CI} is normalized, that is, $\left\langle\Psi^{\mathrm{CI}}\middle|\Psi^{\mathrm{CI}}\right\rangle = 1$. Using the standard technique of Lagrange undetermined multipliers, we obtain the following matrix equation:

$$\mathbf{Hc} = E\mathbf{c}, \qquad (3.156)$$

where **H** is the CI matrix with matrix elements $H_{IJ}^{\mathrm{CI}} = \left\langle\Phi_I\middle|\hat{H}^{el}\middle|\Phi_J\right\rangle$ and **c** is the column vector of CI coefficients. If all the possible excited configurations are included in the CI expansion (full CI), the size of the problems becomes extremely huge and untreatable, even on the best modern computers.

Matrix elements between determinant wave functions are easily calculated using Slater–Condon rules:

1. **Identical determinants**

   We will introduce the designation $\langle ij\|kl\rangle = \langle ij|kl\rangle - \langle ij|lk\rangle$, where

   $$\langle ij|kl\rangle = \int \phi_i^*(\mathbf{r}_1)\phi_j^*(\mathbf{r}_2)\frac{1}{r_{12}}\phi_i(\mathbf{r}_1)\phi_j(\mathbf{r}_2)d\mathbf{r}_1 d\mathbf{r}_2. \qquad (3.157)$$

   Then, for two similar determinant functions, we obtain

   $$\left\langle\Phi_1\middle|\hat{H}\middle|\Phi_1\right\rangle = \sum_m^N \langle m|\hat{h}|m\rangle + \sum_{m>n}^N \langle mn\|mn\rangle. \qquad (3.158)$$

2. **Determinants differing by one spin orbital**

   Let $\langle ij\|kl\rangle = \langle ij|kl\rangle - \langle ij|lk\rangle$,

   $$\langle ij|kl\rangle = \int \phi_i^*(\mathbf{r}_1)\phi_j^*(\mathbf{r}_2)\frac{1}{r_{12}}\phi_i(\mathbf{r}_1)\phi_j(\mathbf{r}_2)d\mathbf{r}_1 d\mathbf{r}_2.$$

   Consider two determinant functions differing by only one spin orbital

   $|\Phi_1\rangle = |...mn...\rangle$ and $|\Phi_2\rangle = |...pn...\rangle$. Then,

$$\langle\Phi_1|\hat{H}|\Phi_2\rangle = \sum_m^N \langle m|\hat{h}|p\rangle + \sum_n^N \langle mn||pn\rangle . \tag{3.159}$$

Consider the matrix element between the ground and singly excited configurations

$\langle\Phi_0|\hat{H}|\Phi_a^r\rangle = \sum_m^N \langle a|\hat{h}|r\rangle + \sum_n^N \langle an||rn\rangle = \langle a|\hat{F}|r\rangle$, where $\hat{F}$ is the HF operator. If $|a\rangle$ is an occupied HF orbital, and $|r\rangle$ is a vacant orbital, then $\langle a|\hat{F}|r\rangle = 0$. That is, in this case

$$\langle\Phi_0|\hat{H}|\Phi_a^r\rangle = 0, \tag{3.160}$$

This result is known as the Brillouin theorem.

3. **Determinant functions differing by two spin orbitals**
   Consider, finally, two determinant functions differing by two spin orbitals $|\Phi_1\rangle = |...mn...\rangle$ and $|\Phi_1\rangle = |...pq...\rangle$. It might be easily shown that $\langle\Phi_1|\hat{H}|\Phi_2\rangle = \langle mn||pq\rangle$.
4. **Determinant functions differing by three and more spin orbitals**
   Let $|\Phi_1\rangle = |...lmn...\rangle$ and $|\Phi_2\rangle = |...pqr...\rangle$. It might be easily shown that the corresponding matrix element is zero.

In practice, the set of considered determinant functions is restricted by the number and type of considered electronic excitations; hence there two main ways to reduce the CI problem. The first way is to reduce the type of excitations that are included in the CI expansion. The hierarchy of such methods includes CI with only singly excited configuration interactions included (CISs) and CI with inclusion of all singly and doubly excited configuration interactions (CISDs).

Of special interest is the CIS approximation, which is widely used for calculations of excited states, especially when the HF wave function is selected as a reference. Note here that because the Hamiltonian contains only one- and two-electron terms, matrix elements between configuration functions differing by more than two orbitals equal zero as a result of orthogonality of one-electron basis functions.

If the HF determinant is selected as the reference, a very important generalized Brillouin theorem is fulfilled. It states that the matrix elements of the total Hamiltonian between the HF ground-state wave function $|\Phi_0\rangle$ and some linear combinations of singly excited configurations equal zero. Therefore, in the CIS approximation, the ground-state wave function remains intact, while the CIS block of the excited state has a dimension of $m \times n$, where m is the number of occupied orbitals and n is the number of virtual orbitals. This theorem is true not only for the HF ground-state wave function but also for any variational solution.

This means that in the CIS approximation the ground-state energy and wave function remain intact, while the solution of the CIS eigenvalue problem gives singly excited states that are orthogonal to the ground state and excitation energies for singly excited states with respect to the HF ground-state energy.

3.5.2 CIS Approximation in Detail

The approach to calculations of excited states using the CIS approximation was proposed in Ref. [30]. The configuration expansion of an approximate solution of the electronic Schrödinger equation is written in the form

$$|\Psi\rangle = c_0|\Phi_0\rangle + \sum_{ia} c_i^a |\Phi_i^a\rangle + \sum_{i<j,a<b} c_{ij}^{ab}|\Phi_{ij}^{ab}\rangle + \dots . \qquad (3.161)$$

Here $|\Phi_0\rangle$ is the reference (HF) configuration, $|\Phi_i^a\rangle$ is a determinant in which a is spin orbital i changed for spin orbital a, and so on.

If we restrict ourselves by the first three expansion terms, we obtain the CISD approximation; if we restrict ourselves by the first two terms, we obtain the CIS approximation [33]. This approximation the following important properties: CIS wave functions are well defined and differentiable; the approximation is applicable to large systems, it is size consistent, and it is variational. The solution are orthogonal to the reference function and to each other.

The following general expression can be obtained for Hamiltonian matrix elements in the CIS approximation:

$$\langle \Phi_i^a | \hat{H} | \Phi_j^b \rangle = E_0 \delta_{ij}\delta_{ab} + F_{ab}\delta_{ij} - F_{ij}\delta_{ab} + \langle aj || ib \rangle . \qquad (3.162)$$

Here, $E_0 = \langle \Phi_0 | \hat{H} | \Phi_0 \rangle$, $F_{pq} = h_{pq} + \sum_{k\in\Phi_0} \langle pk || qk \rangle$, and the standard designation $\langle aj||ib\rangle = \langle aj|ib\rangle - \langle aj|bi\rangle$ is used. In particular, the diagonal matrix element equals

$$\left\langle \Phi_i^a \middle| \hat{H} \middle| \Phi_i^a \right\rangle = E_0 + F_{aa} - F_{ii} - \left\langle ia \| ia \right\rangle , \tag{3.163}$$

and a nondiagonal matrix element equals

$$\left\langle \Phi_i^a \middle| \hat{H} \middle| \Phi_j^b \right\rangle = \left\langle aj \middle| ib \right\rangle - \left\langle aj \middle| bi \right\rangle . \tag{3.164}$$

Equations for CIS wave functions and energies can be presented in a general supermatrix form [31]. The equation for CIS eigenvalues for transition energies $\omega_{\text{CIS}} = E_{\text{CIS}} - E_0$ can be written in the form

$$\sum_{ia} \left\{ \left(\varepsilon_a - \varepsilon_i \right) \delta_{ij} \delta_{ab} + \left\langle ib \| ja \right\rangle \right\} c_i^a = \omega_{\text{CIS}} \sum_{ia} c_i^a \delta_{ij} \delta_{ab} . \tag{3.165}$$

This equation can be written in supermatrix form as

$$\mathbf{AX} = \boldsymbol{\omega}\mathbf{X}, \tag{3.166}$$

where **A** is a matrix designation for the Hamiltonian in the space of singly excited configurations, **ω** is the diagonal matrix of excitation energies, and **X** is the matrix of CIS expansion coefficients. The matrix elements of **A** are given by the expression

$$A_{ia,jb} = \left(\varepsilon_a - \varepsilon_i \right) \delta_{ij} \delta_{ab} + \left\langle ib \| ja \right\rangle , \tag{3.167}$$

and the excitation energies are found by the solution of the secular equation

$$(\mathbf{A} - \boldsymbol{\omega})\mathbf{X} = \mathbf{0}, \tag{3.168}$$

that is, by diagonalization of the matrix **A**.

Calculations of spectra in the CIS approximation give qualitatively acceptable results but systematically overestimate transition energies. A scaling factor of 0.72 was proposed in some works [32, 33] in order to improve this discrepancy, which allowed a quite good agreement between the calculated and experimental transition energies to be obtained.

However, the main disadvantage of this approximation is the neglect of electron correlation. Therefore, excited states that contain a significant contribution of many-electron (two-electron) excitations are poorly described within CIS and are somewhat too high. Because of that, the order of some lowest excited states is not reproduced correctly. It was also shown recently that the CIS approximation strongly and systematically overestimates vertical excitation energies for charge transfer (CT) states (by >1 eV) as compared to non-CT excited states [34].

3.5.3 Complete Active Space SCF Procedure in Brief

The second way of decreasing the amount of calculations in CI is to reduce the set of occupied and virtual orbitals from which and to which excitations are considered. The most widely used approach of this type is the so-called complete active space (CAS) SCF procedure (it is given in detail in Ref. [35]; for a review, see Ref. [36]). In the CAS SCF procedure, all the space of MOs is divided into three subspaces:

- The subspace of doubly occupied (core) MOs, which remain doubly occupied in all the configuration functions included in the CAS procedure
- The subspace of partially occupied active space MOs such that the CAS expansion includes all possible determinants that can be constructed with the participation of active space MOs
- The subspace of virtual MOs that are unoccupied in all the SDs included in the CAS expansion (full CI in this subspace)

The obtained solution of the variational problem depends only on the division of the total MO space into these three mutually orthogonal subspaces, which is fully determined by the three projectors $\hat{P}_1$, $\hat{P}_2$, and $\hat{P}_3$. In their turn, the projectors are fully determined by the sets of the corresponding basis vectors.

It should be specially noted that a truncated CI theory is not size consistent in general. Size consistency means that the energy $E(A + B)$ of a system that consist of two interacting parts A and B tends to the sum of energies of two separate systems $E(A)$ and $E(B)$ when the distance between A and B goes to infinity. However, the CIS procedure is size consistent. This is why it is rather popular in calculations of excited states.

3.5.4 Möller–Plesset Perturbation Theory

A very efficient and widely used way of improving the HF approximation by taking into account electron correlation is Möller–Plesset (MP) perturbation theory [37]. In Section 3.1.5, we did not consider the special form of $\hat{H}$, $\hat{H}_0$, or perturbation operator $\hat{v}$. In the case of the many-electron problem, the form of the Hamiltonian is known

$$\hat{H} = \sum_{i}^{N} h(i) + \sum_{i<j}^{N} r_{ij}^{-1}.$$

The main point of MP perturbation theory is that the HF wave function is used as the zero approximation, and the contribution of electron correlation is taken into account by perturbation theory. In this case, the unperturbed Hamiltonian is simply a sum of one-electron Fockians

$$\hat{H}^{(0)} = \sum_{i}^{N} \hat{f}(i) = \sum_{i}^{N}\left[\hat{h}(i)+\hat{v}^{\mathrm{HF}}(i)\right] = \sum_{i}^{N}\left[\hat{h}(i)+\sum_{p}^{N}\left[\hat{J}_p(i)-\hat{K}_p(i)\right]\right]. \tag{3.169}$$

The perturbation operator $\hat{v}$ is readily obtained as the difference

$$\hat{v} = \hat{H} - \hat{H}^0 = \sum_{i}^{N}\hat{h}(i) + \frac{1}{2}\sum_{i,j}^{N} r_{ij}^{-1} - \sum_{i}^{N}\left[\hat{h}(i)+\hat{v}^{HF}(i)\right]$$

$$= \frac{1}{2}\sum_{i,j}^{N} r_{ij}^{-1} - \sum_{i}^{N}\hat{v}^{HF}(i) = \sum_{i>j}^{N} r_{ij}^{-1} - \sum_{i}^{N}\sum_{p}\left[\hat{J}_p(i)-\hat{K}_p(i)\right]. \tag{3.170}$$

It is convenient, as seen previously (see Section 3.5.1, Eq. 3.153), to expand the perturbed wave functions as a linear combination of excited determinants

$$\left|\Psi_0^{(k)}\right\rangle = \sum_{\mu} C_{\mu}^{(k)}\left|\Psi_{\mu}^{(0)}\right\rangle = C_0^{(k)}\left|\Psi_0^{(0)}\right\rangle + \sum_{S} C_S^{(k)}\left|\Psi_S^{(0)}\right\rangle$$

$$+\sum_{D} C_D^{(k)}\left|\Psi_D^{(0)}\right\rangle + \sum_{T} C_T^{(k)}\left|\Psi_T^{(0)}\right\rangle + \ldots \tag{3.171}$$

Here, we skipped the additional index in the expansion coefficients because we assume that we consider only the ground state. The summation indices *S*, *D*, *T*,... simply mean singly, doubly, triply (and so on) excited configurations.

Because of the nature of the electronic Hamiltonian and the selected expansion basis, a number of simplifications arise: $C_0^{(k)} = 0$ for all $k \neq 0$ by virtue of the intermediate normalization (see Eq. 3.11,

Section 3.1.5). The two-electron nature of operators $\hat{H}$ and $\hat{V}$ leads to truncation of the expansion, while the Brillouin theorem must also be fulfilled, which also simplify all the expressions (see the previous section).

Let us now summarize the starting equations.

The electronic Hamiltonian has the form $\hat{H} = \sum_{i}^{N} \hat{h}(i) + \sum_{i<j}^{N} r_{ij}^{-1}$.

The unperturbed Hamiltonian is $\hat{H}^{(0)} = \sum_{i}^{N} \hat{f}(i) = \sum_{i}^{N} \left[\hat{h}(i) + \hat{v}^{\mathrm{HF}}(i) \right]$.

The perturbation is $\hat{H} - \hat{H}^{(0)} = \sum_{i<j}^{N} r_{ij}^{-1} - \sum_{i}^{N} \hat{v}^{\mathrm{HF}}(i)$.

The expansion over the basis of excited HF determinants

$$
\begin{aligned}
\left|\Psi_0^{(k)}\right\rangle &= \sum_{a;r} C_a^{r(k)} \left|\Psi_a^r\right\rangle \\
&+ \sum_{a>b;r>s} C_{ab}^{rs(k)} \left|\Psi_{ab}^{rs}\right\rangle + \sum_{a>b>c;r>s>t} C_{abc}^{rst(k)} \left|\Psi_{abc}^{rst}\right\rangle + \ldots
\end{aligned}
\tag{3.172}
$$

Now, we find perturbation corrections to the (ground-state) energy with the use of equations of Rayleigh–Schrödinger perturbation theory (see Section 3.1.5). Note here that MP perturbation theory is also a variant of Rayleigh–Schrödinger perturbation theory, in which the zero Hamiltonian $\hat{H}_0$ is a sum of HF operators $\hat{F}_i$ over all electrons $\hat{H}_0^{\mathrm{MP}} = \hat{F} = \sum_i \hat{F}_i$.

We present the wave function of the zero order as an antisymmetrized product of spin orbitals $\left|\Psi_0^{(0)}\right\rangle = \left|\chi_1(x_1)\chi_2(x_2)\ldots\chi_N(x_N)\right\rangle$, where x_i are the sets of Cartesian coordinates of the ith electron. The expression for the zero approximation for the ground-state energy takes the form

$$
\begin{aligned}
E_0^{(0)} &= \left\langle \Psi_0^{(0)} \middle| \hat{H}^{(0)} \middle| \Psi_0^{(0)} \right\rangle \\
&= \sum_{i}^{N} \left[\left\langle \Psi_0^{(0)} \middle| \hat{h}(i) \middle| \Psi_0^{(0)} \right\rangle + \left\langle \Psi_0^{(0)} \middle| \hat{v}^{\mathrm{HF}}(i) \middle| \Psi_0^{(0)} \right\rangle \right]
\end{aligned}
$$

$$= \sum_{b}^{occ}\left[\langle b|h|b\rangle + \sum_{c}^{occ}\langle bc|bc\rangle - \langle bc|cb\rangle\right]$$

$$= \sum_{b}^{occ}\left[\langle b|h|b\rangle + \sum_{c}^{occ}\langle bc||bc\rangle\right] = \sum_{b}^{occ}\varepsilon_b$$

The first-order correction can be written as follows:

$$E_0^{(1)} = \left\langle \Psi_0^{(0)} \middle| \hat{v} \middle| \Psi_0^{(0)} \right\rangle$$

$$= \tfrac{1}{2}\sum_{i;j}^{N}\left[\left\langle \Psi_0^{(0)} \middle| r_{ij}^{-1} \middle| \Psi_0^{(0)} \right\rangle - \sum_{i}^{N}\left\langle \Psi_0^{(0)} \middle| \hat{v}^{\mathrm{HF}}(i) \middle| \Psi_0^{(0)} \right\rangle\right]$$

$$= \tfrac{1}{2}\sum_{b;c}^{occ}\langle bc||bc\rangle - \sum_{b;c}^{occ}\langle bc||bc\rangle$$

$$= -\tfrac{1}{2}\sum_{b;c}^{occ}\langle bc||bc\rangle$$

and

$$E^{\mathrm{HF}} = \sum_{a}^{occ}\varepsilon_a - \sum_{a>b}^{occ}\langle ab||ab\rangle = E_0^{(0)} + E_0^{(1)}$$

The second-order correction in Rayleigh–Schrödinger perturbation theory is written in the form

$$E_0^{(2)} = \sum_{\mu \neq n}\frac{\left|\left\langle \Psi_\mu^{(0)} \middle| \hat{v} \middle| \Psi_0^{(0)} \right\rangle\right|^2}{E_0^{(0)} - E_\mu^{(0)}}.$$

It is necessary to determine which excited determinants interact with the ground-state determinant. The contribution of the ground state into the expansion is zero by virtue of the intermediate normalization (see Eq. 3.1.11, Section 3.1.5). Singly excited determinants do also not make a contribution to the second-order perturbation correction to the energy according to the Brillouin theorem

$$\left\langle \Psi_0^{(0)} \middle| \hat{v} \middle| \Psi_a^r \right\rangle = \sum_{i>j}^{N}\left\langle \Psi_0^{(0)} \middle| \hat{r}_{ij}^{-1} \middle| \Psi_a^r \right\rangle - \sum_{i}^{N}\left\langle \Psi_0^{(0)} \middle| \hat{v}^{\mathrm{HF}} \middle| \Psi_a^r \right\rangle$$

$$= \sum_{b}^{N}\langle ab||rb\rangle - \sum_{b}^{N}\langle ab||rb\rangle = 0.$$

The doubly excited determinants can interact with the ground (reference) determinant $\left\langle \Psi_0^{(0)} \middle| \hat{v} \middle| \Psi_{ab}^{rs} \right\rangle = \sum_{i>j}^{N} \left\langle \Psi_0^{(0)} \middle| \hat{r}_{ij}^{-1} \middle| \Psi_{ab}^{rs} \right\rangle - \sum_{i}^{N} \left\langle \Psi_0^{(0)} \middle| \hat{v}^{\mathrm{HF}} \middle| \Psi_{ab}^{rs} \right\rangle = \langle ab \| rs \rangle$.

Because of the two-particle character of the interaction operator $\hat{v}$, all excitations of higher order cannot interact with the ground determinant. Summing over all possible doubly excited configurations, we can reduce the equation for the second-order correction to the following form:

$$E_0^{(2)} = \sum_{a>b;r>s} \frac{\left|\langle ab \| rs \rangle\right|^2}{E_0^{(0)} - \left(E_0^{(0)} + \varepsilon_r - \varepsilon_a + \varepsilon_s - \varepsilon_b\right)} = \sum_{a>b;r>s} \frac{\left|\langle ab \| rs \rangle\right|^2}{E_{ab}^{rs}} \tag{3.173}$$

Thus far, we considered the case of a spin-unrestricted approach. In the case of closed-shell systems, the equation for the second-order correction is even more simplified:

$$E_0^{(2)} = 2\sum_{abrs}^{\frac{N}{2}} \frac{\langle ab|rs\rangle\langle rs|ab\rangle}{\varepsilon_a + \varepsilon_b + \varepsilon_r + \varepsilon_s} - \sum_{abrs}^{\frac{N}{2}} \frac{\langle ab|rs\rangle\langle rs|ba\rangle}{\varepsilon_a + \varepsilon_b + \varepsilon_r + \varepsilon_s} \tag{3.174}$$

Here, a, b, r, and s are indices of spatial orbitals.

The derivation of the third-order correction and higher is very tedious. We will give here only the final result. Let us first redetermine the expression for the matrix elements of the interaction Hamiltonian by subtracting the energy of the zero approximation from the interaction operator

$$\hat{v}_{\mu,\nu} = \left\langle \Psi_\mu^{(0)} \middle| \left(\hat{v} - E_0^{(0)}\right) \middle| \Psi_\nu^{(0)} \right\rangle.$$

Then, the final equation for the third-order correction will take the form

$$E_0^{(3)} = \sum_{ij}^{D} \frac{\hat{v}_{0,i}\left(\hat{v}_{i,j} - \hat{v}_{0,0}\delta_{i,j}\right)\hat{v}_{j,0}}{\left(E_0^{(0)} - E_i\right)\left(E_0^{(0)} - E_j\right)}. \tag{3.175}$$

Here the sum runs over all the possible doubly excited determinants. Finally, we obtain

$$E^{\mathrm{MP3}} = E_0^{(0)} + E_0^{(1)} + E_0^{(2)} + E_0^{(3)} = E^{\mathrm{HF}} + E_0^{(2)} + E_0^{(3)}.$$

Similarly, we will give only the final equation for the fourth-order correction

$$E_0^{(4)} = \sum_{ij}^{D} \frac{\hat{v}_{0,i}\hat{v}_{i,0}\hat{v}_{0,j}\hat{v}_{j,0}}{\left(E_0^{(0)} - E_i\right)\left(E_0^{(0)} - E_j\right)^2} +$$

$$\sum_{ij}^{D}\sum_{j}^{SDTQ} \frac{\hat{v}_{0,i}\left(\hat{v}_{i,j} - \hat{v}_{0,0}\delta_{i,j}\right)\left(\hat{v}_{j,k} - \hat{v}_{0,0}\delta_{j,k}\right)\hat{v}_{j,0}}{\left(E_0^{(0)} - E_i\right)\left(E_0^{(0)} - E_j\right)\left(E_0^{(0)} - E_k\right)}. \qquad (3.176)$$

The sum over j runs over all singly, doubly, triply, and fourthly excited configurations.

Let us now briefly summarize the above discussion of MP perturbation theory in comparison with CI. As distinct from configuration interaction, MP perturbation theory is not a variational approach, and the MP energy is not an upper bound to the exact energy (i.e., the exact energy is not obligatory lower than the MP energy). CI methods are variational, and the energy predicted by CI is the upper bound to the exact energy. MP perturbation theory is both size consistent and size extensive. We discussed already size consistency, which means that $E(A + B) = E(A) + E(B)$ for two noninteracting systems A and B in the limit of infinitely large distances. Size extensivity means linear scalability of the method at any distances. However, restricted or somehow truncated CI is neither size consistent nor size extensive. On the contrary, full CI is variational, size extensive, and size consistent.

3.5.5 Coupled-Cluster Approximation

The coupled-cluster (CC) approximation is another approach to taking into account electron correlation, which was developed to avoid some deficiencies of perturbative schemes. The history of its development is well described in Ref. [38]. The CC approach is based on the diagrammatic representation of many-body perturbation theory and the use of an important result obtained by Hubbard [39], who showed that only linked (coupled) diagrams contribute to the energy. This result gave the method its name.

The essence of the method is that it uses an exponential expansion of the wave function of the form $|\Psi\rangle = e^{\hat{T}}|\Phi_0\rangle$, where

$\hat{T}$ is the cluster operator, which generates a linear combination of excited determinants from the reference wave function = $|\Phi_0\rangle = |\phi_q \ldots \phi_s\rangle$, where functions $\{\phi_q\}$ form an orthonormalized system (e.g., HF one-electron functions). The method is best described in terms of second quantization operators, and we briefly describe here this technique.

3.5.5.1 Secondary quantization

Let us introduce secondary quantization operators, creation operator $a_p^\dagger$ such that

$$a_p^\dagger \left|\phi_q \ldots \phi_s\right\rangle = \left|\phi_p \phi_q \ldots \phi_s\right\rangle,$$

and annihilation operator a_p such that

$$a_p \left|\phi_p \phi_q \ldots \phi_s\right\rangle = \left|\phi_q \ldots \phi_s\right\rangle.$$

Operators $a_p^\dagger$ and a_p are a Hermitian conjugate. An arbitrary SD can be presented as a chain of creation operators acting on the *vacuum* state

$$a_p^\dagger a_q^\dagger \ldots a_s^\dagger \left|0\right\rangle = \left|\phi_p \ldots \phi_s\right\rangle, \text{ while } a_p|0\rangle = 0.$$

Because $a_p^\dagger a_q^\dagger \left|0\right\rangle = \left|\phi_p \phi_q\right\rangle = -\left|\phi_q \phi_p\right\rangle = -a_q^\dagger a_p^\dagger \left|0\right\rangle$, creation and annihilation operators obey anticommutation relations $a_p^\dagger a_q^\dagger + a_q^\dagger a_p^\dagger = 0$, $a_p a_q + a_q a_p = 0$ $a_p^\dagger a_q + a_q a_p^\dagger = \delta_{pq}$.

With the use of the second quantization operators, we introduce n-orbital *cluster* operators: one-orbital operator $\hat{t}_i \equiv \sum_{a>b} t_i^a a_a^\dagger a_i$ and two-orbital operator $\hat{t}_{ij} \equiv \sum_{a>b} t_{ij}^{ab} a_a^\dagger a_b^\dagger a_j a_i$. In these equations, creation operators act only on the virtual orbitals, while annihilation operators act only on the occupied orbitals. Therefore, all the creation and annihilation operators in equations for t_i and t_{ij} obey anticommutation relations $a_a^\dagger a_i + a_i a_a^\dagger = \delta_{ia} = 0$. Because operators t_i and t_{ij} contain an even number of second quantization operators, they exactly commute. We define now full one- and two-orbital cluster operators

$$\hat{T}_1 \equiv \sum_i \hat{t}_i = \sum_{ia} t_i^a a_a^\dagger a_i \text{ and } \hat{T}_2 \equiv \tfrac{1}{2}\sum_{ij} \hat{t}_{ij} = \tfrac{1}{4}\sum_{ijab} t_{ij}^{ab} a_a^\dagger a_b^\dagger a_j a_i .$$

3.5.5.2 Taking into account pair correlations in the wave function

More generally, we may define an n-orbital cluster operator as $\hat{T}_n \equiv \left(\frac{1}{n!}\right)^2 \sum_{ij...ab...}^{n} \hat{t}_{ij...}^{ab...} a_a^\dagger a_b^\dagger ... a_j a_i$.

As an example, consider the case of the four-electron system $\Phi_0 = |\phi_i(x_1)\phi_j(x_2)\phi_k(x_3)\phi_l(x_4)\rangle$.

A function of two variables can be expanded into a series in products of one-electron functions $f(x_1,x_2) = \sum_{p>q} c_{pq}\phi_p(x_1)\phi_q(x_2)$.

Correlations of any pair of electrons can be taken into account with the use of two-particle cluster functions, $f_{ij}(x_m,x_n) = \sum_{a>b} t_{ij}^{ab}\phi_a(x_m)$ $\phi_b(x_n)$, where t_{ij}^{ab} are cluster coefficients, whose values are determined by solving the electronic Schrödinger equation. The wave function can be improved if such two-particle functions are included into the equation for Φ_0

$$\Psi = \left|\left[\phi_i(x_1)\phi_j(x_2) + f_{ij}(x_1,x_2)\right]\phi_k(x_3)\phi_l(x_4)\right\rangle$$

or

$$\Psi = \Phi_0 + \sum_{a>b} t_{ij}^{ab}\left|\phi_a(x_1)\phi_b(x_2)\phi_k(x_3)\phi_l(x_4)\right\rangle.$$

Finally, a four-electron function that will include only all one- and two-orbital clusters can be written as

$$\begin{aligned}
\Psi = &\left|\phi_i\phi_j\phi_k\phi_l\right\rangle + \left|f_i\phi_j\phi_k\phi_l\right\rangle + \left|\phi_i f_j\phi_k\phi_l\right\rangle + \left|\phi_i\phi_j f_k\phi_l\right\rangle + \left|\phi_i\phi_j\phi_k f_l\right\rangle \\
&+ \left|f_i f_j\phi_k\phi_l\right\rangle + \left|f_i\phi_j f_k\phi_l\right\rangle + \left|f_i\phi_j\phi_k f_l\right\rangle + \left|\phi_i f_j f_k\phi_l\right\rangle + \left|\phi_i f_j\phi_k f_l\right\rangle \\
&+ \left|\phi_i\phi_j f_k f_l\right\rangle + \left|f_i f_j f_k\phi_l\right\rangle + \left|f_i f_j\phi_k f_l\right\rangle + \left|\phi_i f_j f_k\phi_l\right\rangle + \left|\phi_i f_j\phi_k f_l\right\rangle \\
&+ \left|f_{ij}\phi_k\phi_l\right\rangle - \left|f_{ik}\phi_j\phi_l\right\rangle + \left|f_{il}\phi_j\phi_k\right\rangle + \left|\phi_i f_{jk}\phi_l\right\rangle - \left|\phi_i f_{jl}\phi_k\right\rangle \\
&+ \left|\phi_i\phi_j f_{kl}\right\rangle + \left|f_{ij} f_{kl}\right\rangle - \left|f_{ik} f_{jl}\right\rangle + \left|f_{il} f_{jk}\right\rangle + \left|f_i f_j f_k f_l\right\rangle \\
&+ \left|f_{ij} f_k\phi_l\right\rangle + \left|f_{ij}\phi_k f_l\right\rangle + \left|f_{ij} f_k f_l\right\rangle - \left|f_{ik} f_j\phi_l\right\rangle - \left|f_{ik}\phi_j f_l\right\rangle - \left|f_{ik} f_l f_j\right\rangle \\
&+ \left|f_{ij} f_j\phi_k\right\rangle + \left|f_{il}\phi_j f_k\right\rangle + \left|f_{il} f_j f_k\right\rangle + \left|f_i f_{jk}\phi_l\right\rangle + \left|\phi_i f_{jk} f_l\right\rangle + \left|f_i f_{jk} f_l\right\rangle \\
&- \left|f_i f_{jl}\phi_k\right\rangle - \left|\phi_i f_{jl} f_k\right\rangle - \left|f_i f_{jl} f_k\right\rangle + \left|f_i\phi_l f_{kl}\right\rangle + \left|\phi_i f_j f_{kl}\right\rangle + \left|f_i f_j f_{kl}\right\rangle.
\end{aligned}$$

With the use of cluster operators that we defined before, it is possible to present the four-electron function in a rather compact form as

$$\Psi = \left(1 + \sum_{i} \hat{t}_i + \tfrac{1}{2} \sum_{ij} \hat{t}_i \hat{t}_j + \tfrac{1}{6} \sum_{ijk} \hat{t}_i \hat{t}_j \hat{t}_k + \tfrac{1}{2} \sum_{ij} \hat{t}_{ij} + \tfrac{1}{8} \sum_{ijkl} \hat{t}_{ij} \hat{t}_{kl} \right.$$

$$\left. + \tfrac{1}{24} \sum_{ijkl} \hat{t}_i \hat{t}_j \hat{t}_k \hat{t}_l + \tfrac{1}{2} \sum_{ijk} \hat{t}_{ij} \hat{t}_k \hat{t}_l + \tfrac{1}{4} \sum_{ijkl} \hat{t}_{ij} \hat{t}_k \hat{t}_l \right) \Phi .$$

This expression can be simplified with the use of full cluster operators $\hat{T}_1$, $\hat{T}_2$, and $\hat{T}_n$ defined before:

$$\Psi = \left(1 + \hat{T}_1 + \tfrac{1}{2!} \hat{T}_1^2 + \tfrac{1}{3!} \hat{T}_1^3 + \hat{T}_2 + \tfrac{1}{2!} \hat{T}_2{}^2 + \tfrac{1}{4!} \hat{T}_1{}^4 + \hat{T}_2 \hat{T}_1 + \tfrac{1}{2!} \hat{T}_2 \hat{T}_1{}^2 \right) \Phi_0 \tag{3.177}$$

Because $\hat{T}_1$ and $\hat{T}_2$ commute, the above expression is an expansion of an exponent

$$\Psi = e^{\hat{T}_1 + \hat{T}_2} \Phi_0 \equiv e^{\hat{T}} \Phi_0 . \tag{3.178}$$

3.5.5.3 Size consistency

Cluster operators are often called excitation operators. Truncation of cluster operators at a certain level leads to a hierarchy of CC methods (CCSD, CCSDT, etc.).

Let a system consists of two, infinitely separated uncoupled fragments X and Y. Then, cluster operators can be divided into components, so $\hat{T} = \hat{T}_X + \hat{T}_Y$. Hence, $\Psi_{\mathrm{CC}} = e^{\hat{T}} \Phi_0 = e^{\hat{T}_X + \hat{T}_Y} \Phi_0 = e^{\hat{T}_X} e^{\hat{T}_Y} \Phi_0$.

Because the wave function Φ_0 is also factorized, we obtain as a result $E_{\mathrm{CC}} = E_{\mathrm{CC}}^X + E_{\mathrm{CC}}^Y$. This is not so in the case of the CI approach. Actually

$\Psi_{\mathrm{CI}} = \left(1 + \hat{C} \right) \Phi_0 = \left(1 + \hat{C}_X + \hat{C}_Y \right) \Phi_0$, and as a result $E_{\mathrm{CI}} \neq E_{\mathrm{CI}}^X + E_{\mathrm{CI}}^Y$, if only the excitation operator is not truncated.

3.5.5.4 Formal solution of the problem of coupled-cluster theory

We start from the Schrödinger equation

$$\hat{H}|\Psi\rangle = E|\Psi\rangle$$

and substitute the wave function in the form $\Psi_{CC} = e^{\hat{T}}\Phi_0$ in this equation. Then, we obtain

$$\hat{H}e^{\hat{T}}|\Phi_0\rangle = Ee^{\hat{T}}|\Phi_0\rangle. \tag{3.179}$$

If Eq. 3.157 is multiplied by Φ_0 from the left, we will obtain the expression for the energy

$$\langle\Phi_0|\hat{H}e^{\hat{T}}|\Phi_0\rangle = E\langle\Phi_0|e^{\hat{T}}|\Phi_0\rangle = E. \tag{3.180}$$

(it is supposed that the so-called intermediate normalization $\langle\Phi_0|\Psi_{CC}\rangle = 1$ is used, as well as in the case of perturbation theory; see Section 3.1.5, Eq. 3.11). Multiplying Eq. 3.157 by excited determinants from the left, one can obtain equations for all amplitudes. These equations are formally exact; they are nonlinear and energy dependent.

3.5.5.5 Derivation of equations of coupled-cluster theory

An exponential operator can be expanded into a power series

$$e^{\hat{T}} = 1 + \hat{T} + \frac{\hat{T}^2}{2!} + \frac{\hat{T}^3}{3!} + \dots. \tag{3.181}$$

If this expansion is substituted into the equation for energy, we will obtain $\langle\Phi_0|\hat{H}\left(1 + \hat{T} + \frac{\hat{T}^2}{2!} + \frac{\hat{T}^3}{3!} + \dots\right)|\Phi_0\rangle = E$, and

$$\langle\Phi_0|\hat{H}|\Phi_0\rangle + \langle\Phi_0|\hat{H}\hat{T}|\Phi_0\rangle + \langle\Phi_0|\hat{H}\frac{\hat{T}^2}{2!}|\Phi_0\rangle + \langle\Phi_0|\hat{H}\frac{\hat{T}^3}{3!}|\Phi_0\rangle + \dots = E. \tag{3.182}$$

The third and the next terms will give a zero contribution, so

$$\langle\Phi_0|\hat{H}|\Phi_0\rangle + \langle\Phi_0|\hat{H}\hat{T}|\Phi_0\rangle + \langle\Phi_0|\hat{H}\frac{\hat{T}^2}{2!}|\Phi_0\rangle = E.$$

The usual form of the method of CCs is obtained by the multiplication of the Schrödinger equation by $e^{-\hat{T}}$ from the left $e^{-\hat{T}}\hat{H}e^{\hat{T}}|\Phi_0\rangle = E|\Phi_0\rangle$, which results in the equations for the energy $\langle\Phi_0|e^{-\hat{T}}\hat{H}e^{\hat{T}}|\Phi_0\rangle = E$ and for the amplitudes $\langle\Phi_{ij\dots}^{ab\dots}|e^{-\hat{T}}\hat{H}e^{\hat{T}}|\Phi_0\rangle = 0$.

The use of the Baker–Campbell–Hausdorff formula for $e^{-\hat{T}}\hat{H}e^{\hat{T}}$ gives $e^{-\hat{T}}\hat{H}e^{\hat{T}} = \hat{H} + \left[\hat{H},\hat{T}\right] + \frac{1}{2!}\left[\left[\hat{H},\hat{T}\right],\hat{T}\right] + \ldots$, which allows the infinite expansion to be cut off in a natural way. Then, the presentation of the Hamiltonian $\hat{H}$ in terms of operators $a_p^{\dagger}$, a_p

$$\hat{H} = \sum_{pq} h_{pq} a_p^{\dagger} a_q + \tfrac{1}{4}\sum_{pqrs} \langle pq \| rs \rangle a_p^{\dagger} a_q^{\dagger} a_s a_r \qquad (3.183)$$

is used for the construction of a computational scheme based on the so-called method of equations of motion; see Ref. [40].

3.5.6 Concluding Remarks

In orbital methods, that is, methods based on the LCAO approximation, matrix elements of all operators over MOs are calculated through one-electron and two-electron integrals over the AOs, used in the LCAO expansion. Two-electron integrals in terms of basis set AOs χ_m (see Eq. 3.133) form a very large array

$$g_{\kappa\lambda;\mu\nu} = \left\langle \chi_\kappa \chi_\lambda \left| r_{12}^{-1} \right| \chi_\mu \chi_\nu \right\rangle = \int \chi_\kappa(r_1)^* \chi_\lambda(r_2)^* \chi_\mu(r_1) \chi_\nu(r_2) dr_1 dr_2 , \qquad (3.184)$$

whose size is $\sim \frac{N^4}{8}$. The various designations and other details for these integrals are thoroughly discussed in Ref. [3]. The main computational problems in all orbital methods are associated with the necessity of running over the array of two-electron integrals many times.

Note here that the effect of double substitutions, which are absent in CIS excited states, can be included using a second-order perturbation expansion of the CC excited-state method, restricted to single and double substitutions, in a series in which CIS is of the zeroth order, and the first-order correction vanishes. This method was developed for the case of closed shells in Ref. [41] and in Ref. [42] for excited states of radicals. CIS(D) excitation energies are size consistent, and the required computational resources are not too high and scale as the fifth power of molecular size, as well as in second-order MP theory.

We will conclude this section by a table of comparative characteristics of different quantum-chemical methods:

Method	Scaling	Iterative
SCF	$\propto N^4$	Yes
MP2	$\propto N^5$	No
MP3	$\propto N^6$	No
MP4	$\propto N^7$	No
CISD	$\propto N^6$	Yes
CCSD	$\propto N^6$	Yes

3.6 Basics of Density Functional Theory

3.6.1 Introduction: Hohenberg–Kohn Theorems

Density functional theory (DFT) is now the main quantum-mechanical approach that is used in atomistic multiscale simulation of various molecular and extended systems. The important computational advantage of DFT is that it avoids direct calculations of the huge array of two-electron integrals required for the calculations of HFR equations or matrix elements between configurations in post-HF methods (MP2–MP4, CCSD, CISD, etc.). For more details about DFT, we may recommend some fundamental books [43, 44].

It is generally accepted that DFT is methodologically based on the famous Hohenberg–Kohn theorems [45].

The first Hohenberg–Kohn theorem (HK-I) states that one-to-one correspondence exists between the external (local one-electron) potential $v(r)$, the electron density function $\rho(r)$, and the wave function of the ground state: $v(r) \Leftrightarrow \rho(r) \Leftrightarrow \Psi_0$.

Proof: Consider the electron density function for a nondegenerate ground state, $\rho(r)$. First, it determines the total number of electrons in the system under consideration (because $\int \rho(r)dr = N$). However, $\rho(r)$ determines $v(r)$ because otherwise we would have two different Hamiltonians $\hat{H}$ and $\hat{H}'$, differing in potentials $v(r)$ and $v'(r)$, for which the ground-state densities would be different. Because $\hat{H}\Psi = E_0\Psi$ and $\hat{H}'\Psi' = E_0'\Psi'$, then Ψ

$\neq \Psi'$. Otherwise, we have have $\hat{H}\Psi = E_0\Psi$ and $\hat{H}'\Psi = E_0'\Psi$, that is, $(\hat{H} - \hat{H}')\Psi = (v - v')\Psi = (E_0 - E_0')\Psi$, and $v - v' = E_0 - E'_0$. From here, it follows that the two potentials, $v(r)$ and $v'(r)$, differ only by a constant. Moreover, using the variational principle, we have

$$E_0 < \langle \Psi' | \hat{H} | \Psi' \rangle = \langle \Psi' | \hat{H} | \Psi' \rangle + \langle \Psi' | \hat{H} - \hat{H}' | \Psi' \rangle$$

$$= E_0' + \int \rho(r)(v(r) - v'(r))dr,$$

$$E_0' < \langle \Psi | \hat{H}' | \Psi \rangle = \langle \Psi | \hat{H} | \Psi \rangle + \langle \Psi | \hat{H} - \hat{H}' | \Psi' \rangle$$

$$= E_0 - \int \rho(r)(v(r) - v'(r))dr.$$

Summing up these two inequalities, we obtain that $E_0 + E_0' < E_0' + E_0$, which is a contradiction. This proves the first Hohenberg–Kohn theorem.

Hence, $\rho(r)$ uniquely determines N and $v(r)$ and, hence, it determines all the properties of the ground state of the system: kinetic energy $T[\rho(r)]$, potential energy $V[\rho(r)]$, and total energy $E[\rho(r)]$. So, we can write

$E_v[\rho(r)] = T[\rho] + V_{ne}[\rho] + V_{ee}[\rho] = \int \rho(r)v(r)dr + F_{HK}[\rho]$, where $F_{HK}[\rho]$is the Hohenberg–Kohn functional defined as

$$F_{HK}[\rho] = T[\rho] + V_{ee}[\rho]. \tag{3.185}$$

Then, we can also assume that

$$V_{ee}[\rho] + E_{XC}[\rho], \tag{3.186}$$

where $J[\rho] = \frac{1}{2}\int \frac{\rho(r_1)\rho(r_2)}{r_{12}} dr_1 dr_2$, while $E_{XC}[\rho]$ is called the exchange correlation functional.

The second Hohenberg–Kohn theorem (HK-II) states that the density functional obeys a variational principle, namely for any trial density $\tilde{\rho}(r)$ such that $\tilde{\rho}(r) \geq 0$ and $\int \tilde{\rho}(r)dr = N$, $E_0 \leq E_v[\tilde{\rho}(r)]$, where $E_v[\tilde{\rho}(r)]$ is the energy functional defined before.

Proof: The first Hohenberg–Kohn theorem states that the density $\tilde{\rho}(r)$ determines the corresponding potential $\tilde{v}(r)$ and, hence, the Hamiltonian $\tilde{\hat{H}}$ and wave function $\tilde{\Psi}$. Considering wave function $\tilde{\Psi}$ as a trial wave function for the Hamiltonian with potential $v(r)$, we obtain the inequality

$$\langle\tilde{\Psi}|\hat{H}|\tilde{\Psi}\rangle = \int \tilde{\rho}(r)v(r)dr + F_{\text{HK}}[\tilde{\rho}] = E_v[\tilde{\rho}] \geq E_v[\rho].$$

This inequality represents the formulation of the variational principle for the density functional, which proves the second Hohenberg–Kohn theorem.

The variational principle assumes the differentiability of $E_v[\rho]$ and requires that the ground-state density obey the stationarity condition

$$\delta\left\{E_v[\rho] - \mu\left[\int \rho(r)dr - N\right]\right\} = 0. \tag{3.187}$$

The stationarity condition Eq. 3.187 leads to the common Euler–Lagrange equations

$\mu = \dfrac{\delta E_v[\rho]}{\delta\rho(r)} = v(r) + \dfrac{\delta F_{\text{HK}}[\rho]}{\delta\rho(r)}$, where μ has the meaning of the chemical potential of the system.

Remember here that according to Eqs. 3.185 and 3.186,

$$F_{\text{HK}}[\rho] = T[\rho] + V_{\text{ee}}[\rho] = T[\rho] + J[\rho] + E_{XC}[\rho]. \tag{3.188}$$

If the exact form of $F_{\text{HK}}[\rho]$was known explicitly, the variational equation (Eq. 3.187) would be an exact equation for the ground-state electron density.

However, it is important that the functional $F_{\text{HK}}[\rho]$ not depend on the potential $v(r)$ and, hence, is a universal functional of density $\rho(r)$. Correspondingly, functionals $T[\rho]$and $E_{XC}[\rho]$are also universal functional of density $\rho(r)$.

There is also the problem of so-called v-representability, which was first discussed by Kohn [46], that is, the existence of (physically reasonable) $\rho(r)$ for any (physically reasonable) $v(r)$. The problem of v-representability can be circumvented in the following way:

Consider the condition $E_0 = \min_{\Psi}\langle\Psi|\hat{T} + \hat{V}_{\text{ee}} + \sum_i^N v(\mathbf{r}_i)|\Psi\rangle$. It is possible to minimize the energy over all wave functions that give the required density $\rho(r)$ and then minimize the energy over all densities

$$E_0 = \min_{\rho}\left(\min_{\Psi\to\rho}\langle\Psi|\hat{T} + \hat{V}_{\text{ee}} + \sum_i^N v(\mathbf{r}_i)|\Psi\rangle\right). \tag{3.189}$$

The term in parentheses, depending on the external potential, explicitly depends on density, and it can be excluded from minimization with respect to Ψ as

$$E_0 = \min_{\rho}\left(\min_{\Psi\to\rho}\langle\Psi|\hat{T}+\hat{V}_{ee}|\Psi\rangle + \int v(\mathbf{r})\rho(\mathbf{r})d\mathbf{r}\right)$$
$$= \min_{\rho}\left(F[\rho] + \int v(\mathbf{r})\rho(\mathbf{r})d\mathbf{r}\right), \quad (3.190)$$

where $F[\rho] = \min_{\Psi\to\rho}\langle\Psi|\hat{T}+\hat{V}_{ee}|\Psi\rangle$.

That is, it is sufficient to consider only all N-representable densities.

Let us remember that $E_v[\rho(r)] = T[\rho] + V_{ne}[\rho] + V_{ee}[\rho] = \int\rho(r)v(r)dr + F_{HK}[\rho]$, where $F_{HK}[\rho] = T[\rho] + V_{ee}[\rho] = T[\rho] + J[\rho] + E_{XC}[\rho]$ (see Eq. 3.188).

While $E_{XC}[\rho]$ is a relatively small correction, for which some quite good approximations have been developed, the value of $T[\rho]$ is comparable with the total energy. Moreover, there is no reliable theoretical expressions for $T[\rho]$, which has been known well since the time when Thomas–Fermi theory was developed.

3.6.2 Kohn–Sham Equations

An ingenious solution of this problem was proposed by Kohn and Sham [47]. Consider a system of *noninteracting* electrons in an external potential $v_{KS}(\mathbf{r})$ such that the corresponding wave function, that is, an SD constructed on N lowest-energy orbitals, gives the same electron density $\rho(\mathbf{r})$ as the exact system of *interacting* electrons in the external potential $v(\mathbf{r})$. The electronic energy can be exactly presented in the form

$$E[\rho] = \int\rho(\mathbf{r})v(\mathbf{r})d\mathbf{r} + F[\rho]. \quad (3.191)$$

Kohn and Sham defined the functional $F[\rho]$ as

$$F[\rho] = T_s[\rho] + J[\rho] + E_{XC}[\rho], \quad (3.192)$$

where $J[\rho]$ is the classical Coulomb repulsion of the density $\rho(\mathbf{r})$ with itself

$$J[\rho] = \frac{1}{2}\iint\frac{\rho(\mathbf{r}_1)\rho(\mathbf{r}_2)}{r_{12}}d\mathbf{r}_1 d\mathbf{r}_2, \quad (3.193)$$

while $T_s[\rho]$ is the kinetic energy of the system of noninteracting electrons.

Because $F[\rho] = T[\rho] + V_{ee}[\rho]$, we obtain

$$E_{XC}[\rho] = T[\rho] - T_s[\rho] + V_{ee}[\rho] - J[\rho], \tag{3.194}$$

and the total energy takes the form

$$E[\rho] = \int \rho(\mathbf{r})v(\mathbf{r})d\mathbf{r} + T_s[\rho] + J[\rho] + E_{XC}[\rho]. \tag{3.195}$$

The key quantity here is the functional $E_{XC}[\rho]$, which is called the exchange correlation functional (see Eq. 3.186). The variational Euler–Lagrange equations take the form

$$\mu = v_{\text{eff}}(\mathbf{r}) + \frac{\delta T_s[\rho]}{\delta \rho(\mathbf{r})}.$$

Here, the effective potential $v_{\text{eff}}(\mathbf{r})$ is defined as

$$v_{\text{eff}}(\mathbf{r}) = v(\mathbf{r}) + \frac{\delta T[\rho]}{\delta \rho(\mathbf{r})} + \frac{\delta V_{ee}[\rho]}{\delta \rho(\mathbf{r})}. \tag{3.196}$$

The Euler–Lagrange equations written above determine the exact electron density of the real system. At the same time, these equations exactly describe the system of noninteracting particles in the external potential $v_{\text{eff}}(\mathbf{r})$. Hence, the electron density of such an artificial system will coincide with the electron density of the real system of interacting particles.

Then, the Kohn–Sham (KS) Hamiltonian will take the form

$$\hat{H}_{\text{KS}} = \sum_i \hat{h}_{\text{KS}}(i) = \sum_i \left(-\frac{1}{2}\nabla^2(i) + v_{\text{KS}}(\mathbf{r}_i) \right). \tag{3.197}$$

This Hamiltonian is a sum of one-electron Hamiltonians $\hat{h}_{\text{KS}}(i)$, which differ only in the electron number "i" and determine the orbital equations

$$\hat{h}_{\text{KS}}(1)\phi_i(1) = \varepsilon_i \phi_i(1). \tag{3.198}$$

The total wave function of the system of noninteracting particles will take an HF form

$$\Psi_{\text{KS}} = |\phi_1(1)\phi_2(2)\ldots\phi_N(N)|, \tag{3.199}$$

and the electron density will be simply defined as

$$\rho_{\text{KS}}(\mathbf{r}) = \sum_{i=1}^{N} \sum_s |\phi_i(\mathbf{r},s)|^2 = \rho^{\text{exact}}(\mathbf{r}). \tag{3.200}$$

Finally, the kinetic energy of the system of noninteracting particles will take a simple form

$$T_s[\rho]=\sum_i^N\langle\phi_i|-\frac{1}{2}\nabla^2|\phi_i\rangle. \tag{3.201}$$

By virtue of the HK-I theorem, the potential $v_{KS}(\mathbf{r})$ must be unique. KS theory is formally exact if the exact exchange correlation potential is used. This potential contains the difference between the exact energy of electron-electron repulsion $V_{ee}[\rho]$ and the Coulomb energy $J[\rho]$, and also between the exact kinetic energy $T[\rho]$ and the kinetic energy of noninteracting particles $T_s[\rho]$.

Let us now give a brief synopsis of the main points of KS theory:

- $v_{KS}(\mathbf{r})$ is a unique local potential associated with the system of noninteracting electrons.
- $v_{KS}(\mathbf{r})$ gives an exact one-electron density $\rho(\mathbf{r})$ (an important distinction from HF theory).
- ε_{HOMO} = –IP (the consequence of the fact that the asymptotic long-range behavior of the electron density of a bound system is governed by the first ionization potential (IP), while the asymptotic long-range behavior of the electron density, in its turn, is determined by the one-electron energy of the highest occupied molecular orbital [HOMO]; see Refs. [48, 49]).
- Energies of unoccupied (vacant) orbitals and ε_{LUMO} in particular correspond solutions in the same potential as occupied MOs; therefore, differences between KS orbital energies can serve as a first approximation to transition energies calculated using time-dependent density functional theory (TDDFT), which will be considered later.

3.6.3 Kohn–Sham Theory: LDA

For the KS approach to be used in practical calculations, it is necessary to find an explicit (at least approximate) expression for the KS potential $v_{KS}(\mathbf{r})$. To do that, one can use a model system for which an exact solution is known. A simplest, exactly solvable system is a uniform electron gas. An approximation to the KS potential based on this model was suggested by the authors of the KS method in Ref. [50]. This approximation, called the local density approximation

(LDA), is applicable for systems with slowly varying electron density $\rho(\mathbf{r})$.

Consider the uniform electron gas in a model in which a large number of electrons, N, is placed in a cube of volume $V = l^3$. The electron density is constant everywhere and is compensated by a uniform positively charged background of the same density. The ground-state electron density can be presented as

$$E[\rho] = T_s[\rho] + \int \rho(\mathbf{r})v(\mathbf{r})d\mathbf{r} + J[\rho] + E_{XC}[\rho] + E_\mathrm{b}, \quad (3.202)$$

where E_b is the electrostatic repulsion energy of the positively charged background. The second, third, and fifth terms can be written as

$$\int \rho(\mathbf{r})v(\mathbf{r})d\mathbf{r} + \frac{1}{2}\iint \frac{\rho(\mathbf{r})\rho(\mathbf{r}')}{|\mathbf{r}-\mathbf{r}'|} + \frac{1}{2}\iint \frac{n(\mathbf{r})n(\mathbf{r}')}{|\mathbf{r}-\mathbf{r}'|} d\mathbf{r}d\mathbf{r}'. \quad (3.203)$$

Making use of the fact that $n(\mathbf{r}) = -\rho(\mathbf{r})$ and also that $v(\mathbf{r}) = -\int \frac{n(\mathbf{r}')}{|\mathbf{r}-\mathbf{r}'|} d\mathbf{r}'$, we obtain that the sum of these three terms equals zero. Therefore, the electronic energy is reduced to the sum

$$E[\rho] = T_s[\rho] + E_{XC}[\rho] = T_s[\rho] + E_X[\rho] + E_C[\rho]. \quad (3.204)$$

The exact exchange energy is determined by the equation

$$E_X[\rho] = -\frac{1}{4}\iint \frac{\rho_1(\mathbf{r}_1,\mathbf{r}_2)}{r_{12}} d\mathbf{r}_1 d\mathbf{r}_2,\ \rho_1(\mathbf{r}_1,\mathbf{r}_2) = 2\sum_i \phi_i(\mathbf{r}_1)\phi_i(\mathbf{r}_2). \quad (3.205)$$

For a uniform electron gas, orbitals are plane waves $\phi_k(\mathbf{r}) = \frac{1}{V^{1/2}} e^{i\mathbf{k}\cdot\mathbf{r}}$; according to periodic boundary conditions, $\mathbf{k} = \frac{2\pi}{l}\mathbf{n}$, $n_x, n_y, n_z = 0, \pm1, \pm2, \ldots$. Therefore, the one-particle density matrix for a uniform electron gas will take the form

$$\rho_1(\mathbf{r}_1,\mathbf{r}_2) = \frac{2}{V} \sum_{\mathbf{k}(\mathrm{occ})} \exp\left[i\mathbf{k}\cdot(\mathbf{r}_1 - \mathbf{r}_2)\right], \quad (3.206)$$

where the sum is taken over all occupied orbitals. Proceeding to the limit from the sum in Eq. 3.206 to an integral $dn_x dn_y dn_z = (l/2\pi)^3\, d\mathbf{k} = (V/8\pi^3)\, d\mathbf{k}$, we obtain

$$\rho_1(\mathbf{r}_1,\mathbf{r}_2) = \frac{1}{4\pi^3} \int \exp\left[i\mathbf{k}\cdot(\mathbf{r}_1 - \mathbf{r}_2)\right] d\mathbf{k}, \quad (3.207)$$

or, in spherical coordinates

$$\rho_1(\mathbf{r}_1,\mathbf{r}_2)=\frac{1}{4\pi^3}\int_0^{k_F}\iint \exp\left[i\mathbf{k}\cdot(\mathbf{r}_1-\mathbf{r}_2)\right]\sin\theta d\theta d\varphi k^2 dk\,. \quad (3.208)$$

The value of the Fermi wave vector $\mathbf{k}_F$ can be obtained by assuming that $\mathbf{r}_1 = \mathbf{r}_2$. Then, $\rho(\mathbf{r})=\frac{\mathbf{k}_F^3}{3\pi^2}$. Introducing coordinates $\mathbf{r} = (\mathbf{r}_1 + \mathbf{r}_2)/2$, $\mathbf{S} = (\mathbf{r}_1 - \mathbf{r}_2)$ and choosing $\mathbf{S}$ directed along axis k_z, we obtain

$$\rho_1(\mathbf{r}_1,\mathbf{r}_2)=\frac{1}{2\pi^2}\int_0^{\mathbf{k}_F}\int_{\theta=0}^{\pi} e^{iks\cos\theta}\sin\theta d\theta k^2 dk\,. \quad (3.209)$$

This expression can be integrated, which gives the final result

$$\rho_1(\mathbf{r}_1,\mathbf{r}_2)=3\rho(\mathbf{r})\frac{\sin t - t\cos t}{t^3},\ t=\mathbf{k}_F s\,. \quad (3.210)$$

Now, we can calculate the exchange energy $E_X[\rho]=-\frac{1}{4}\iint\frac{\rho_1(\mathbf{r}_1,\mathbf{r}_2)}{r_{12}}d\mathbf{r}_1 d\mathbf{r}_2$, which gives the well-known Dirac formula (see, for example, Ref. [50]) for the quantity $E_X[\rho]$

$$E_X[\rho]=C_X\int\rho^{4/3}(\mathbf{r})d\mathbf{r},\ C_X=-\frac{3}{4}(3/\pi)^{1/3}=0.7386\,. \quad (3.211)$$

The kinetic energy of a uniform electron gas in LDA can also be calculated taking into account that $T[\rho]=\int-\frac{1}{2}\nabla^2\rho_1(\mathbf{r}_1,\mathbf{r}_2)\Big|_{\mathbf{r}_2=\mathbf{r}_1}d\mathbf{r}_1$ and that the action of operator ∇^2 is equivalent to multiplication by k^2. This leads to the well-known Thomas–Fermi equation [51] for the kinetic functional

$$T^{\mathrm{TF}}[\rho]=C_F\int\rho^{5/3}(\mathbf{r})d\mathbf{r},\ C_F=\frac{3}{10}\left(3\pi^2\right)^{2/3}=2.8712\,. \quad (3.212)$$

Now, it is remained to calculate the correlation energy for the uniform electron gas model. In the classical work by D. M. Ceperley and B. J. Alder [52], the correlation energy of a uniform electron gas was calculated by the quantum Monte Carlo method. The authors calculated the correlation energy of a uniform electron gas for several values of the Wigner–Seitz radius $r_s=\left(\frac{3}{4\pi}\right)^{1/3}\rho(\mathbf{r})^{-1/3}$

(the radius of a sphere containing one electron) by subtracting the kinetic and exchange energies from the total energy (see Eq. 3.195)

$$E[\rho] = T_s[\rho] + E_{XC}[\rho] = T_s[\rho] + E_X[\rho] + E_C[\rho]. \tag{3.213}$$

Then, using analytical data for the limits of high and low densities, the authors of Ref. [53] obtained the following expression for $\varepsilon_C(r_s)$, called Vosko–Wilk–Nusair (VWN) functional:

$$\varepsilon_C(r_s) = \frac{A}{2}\left(\ln\frac{x^2}{X(x)} + \frac{2b}{Q}\arctan\frac{Q}{2x+b}\right) - \frac{bx_0}{X(x_0)}\left[\ln\frac{(x-x_0)^2}{X(x)} + \frac{2(b+2x_0)}{Q}\arctan\frac{Q}{2x+b}\right], \tag{3.214}$$

where $x = r_s^{1/2}$, $X(x) = x^2 + bx + c$, $Q = (4c - b^2)^{1/2}$, and $A = 0.0621814$, 0.0310907; $x_0 = -0.409286, -0.743294$; $b = 13.0720, 20.1231$; and $c = 42.7198, 101.578$ for $\varepsilon_C^0(r_s)$ and $\varepsilon_C^1(r_s)$; indices 0 and 1 correspond to fully paired and fully unpaired electrons, respectively.

So, summing up all this, we may conclude that in LDA calculations, the Thomas–Fermi equation is used for the kinetic energy functional, the Dirac (Slater) approximation is used for the exchange functional, and the VWN functional is used for the correlation term.

3.6.4 Beyond LDA: Generalized Gradient Approximation

It was found at the early stages of the development of DFT that LDA has many deficiencies, the main source of which is the use of the uniform electron gas model. Among these deficiencies are the problem of the self-interaction correction (the exchange correlation potential, Eq. 3.186, must compensate for the interaction of an electron with itself, which spuriously arises in the Coulomb term) and the problem of asymptotic long-range behavior of the exchange correlation potential (the exchange correlation potential must decrease as r^{-1} as $r \to \infty$).

While the electron density is roughly uniform in extended systems (like crystals), it is strongly nonuniform in molecular systems. This fact explains the success of LDA DTF in solid-state calculations and relative failure in molecular calculations. To take into account the inhomogeneity of the electron density, it was suggested to make the exchange and correlation energies dependent not only on the electron density but also on derivatives of the density and include a dependence of the exchange correlation potential on the gradients of electron density in the exchange correlation potential.

Such methods, in which the gradients of electron density are explicitly included in $E_{XC}[n]$, are known as gradient-corrected or generalized gradient approximation (GGA) methods. A straightforward Taylor expansion does not lead to an improvement over LDA; it actually makes things worse, thus the name "generalized gradient approximation." GGA methods are also sometimes referred to as nonlocal methods, although this is somewhat misleading because the functionals depend only on the density (and derivatives) at a given point, as opposed to, for example, the HF exchange energy. For further discussion of these points, see Refs. [3, 4, 54, 55] and Refs. [56–57].

In accordance with this, it is assumed that

$$E_{XC}^{\mathrm{GGA}}[\rho_\alpha,\rho_\beta]=\int drf\left(\rho_\alpha,\rho_\beta,\vec{\nabla}\rho_\alpha,\vec{\nabla}\rho_\beta\right), \tag{3.215}$$

where ρ_α and ρ_β are densities of electrons with spins α and β. In practice, it is commonly accepted that $E_{XC}^{\mathrm{GGA}}=E_X^{\mathrm{GGA}}+E_C^{\mathrm{GGA}}$. Let us write the exchange part in the form

$$E_X^{\mathrm{GGA}}=E_X^{\mathrm{LDA}}-\sum_\sigma\int F(s_\sigma)\rho_\sigma^{4/3}(\vec{r})d\vec{r}, \tag{3.216}$$

where

$$E_X^{\mathrm{LDA}}[\rho]=C_X\int\rho^{4/3}(\mathbf{r})d\mathbf{r},\ C_X=-\frac{3}{4}(3/\pi)^{1/3}=0.7386,$$

and

$$s_\sigma=\frac{|\nabla\rho_\sigma|}{\rho_\sigma^{4/3}}. \tag{3.217}$$

The energy of electron-electron interaction can be presented as

$$V_{ee}[\rho]=\iint\frac{1}{r_{12}}\rho_2(\mathbf{r}_1,\mathbf{r}_2)d\mathbf{r}_1d\mathbf{r}_2,$$

$$\rho_2(\mathbf{r}_1,\mathbf{r}_2)=\tfrac{1}{2}\rho(\mathbf{r}_1)\rho(\mathbf{r}_2)\left[1+h(\mathbf{r}_1,\mathbf{r}_2)\right], \tag{3.218}$$

where $\rho_2(\mathbf{r}_1,\mathbf{r}_2)$ is a diagonal two-particle density matrix and $h(\mathbf{r}_1,\mathbf{r}_2)$ is a pair correlation function.

The exchange correlation hole is defined as follows:

$$\rho_{xc}(\mathbf{r}_1,\mathbf{r}_2) = r(\mathbf{r}_2)h(\mathbf{r}_1,\mathbf{r}_2). \tag{3.219}$$

From the definition of the two-particle density matrix, we obtain $\int \rho_{xc}(\mathbf{r}_1,\mathbf{r}_2)d\mathbf{r}_2 = -1$, that is, the charge of the exchange correlation hole. This fact corroborates the statement that the exchange correlation potential must decrease as r^{-1} as $r \to \infty$.

Let us express the exchange energy through the exchange potential $\varepsilon_x(\mathbf{r}_1)$

$$E_x = \tfrac{1}{2}\int \rho(\mathbf{r}_1)\varepsilon_x(\mathbf{r}_1)d\mathbf{r}_1\ ,\ \varepsilon_x(\mathbf{r}_1) = \int \frac{\rho_{xc}(\mathbf{r}_1,\mathbf{r}_2)}{r_{12}}d\mathbf{r}_2 \tag{3.220}$$

From this equation, it follows that at $r_1 \to \infty$, $\lim\limits_{x\to\infty} \varepsilon_x(\mathbf{r}_1) = -\dfrac{1}{r_1}$. However, the Slater (Dirac) exchange potential decreases exponentially. Therefore, Axel Becke proposed potentials with a correct asymptotic behavior, selecting the function $F(x)$, where $x = s_\sigma$. In 1986, Becke proposed the expression $F(x) = C_x + \beta\left(x^2/(1+\gamma x^2)\right)$ (potential B86X, [63]), which was further used in the Perdew–Burke–Ernzerhof (1996) (PBE) exchange functional [58], and later in Ref. [59] Becke proposed an exchange potential of the form $F(x) = C_x + \beta x^2/(1 + 6\beta x \,\text{archinh}\, x)$ (B88X potential, $\text{arcsinh}\, x = \ln\left(x + \sqrt{x^2+1}\right)$). Parameter β was determined by fitting to exchange energies of noble gas atoms.

Now, we pass to GGA correlation potentials. In a very important paper, Colle and Salvetti obtained an approximate expression for the correlation energy of a He atom in terms of the HF second-order density matrix [60]. On the basis of this formula, Lee, Yang, and Parr (LYP) obtained an explicit expression (the LYP functional) for the electron density functional [61]. The LYP functional has a rather complicated analytical form. Its great advantage is that it was obtained for an actual correlated wave function of a two-electron system and is not completely connected with the uniform electron gas model. A combination of the Becke exchange potential and the LYP correlation potential is called the BLYP functional.

It should be noted that GGA functionals determine exchange correlation potentials that are local operators in the sense of their action onto wave functions; that is, these are operators of multiplication by a function. The further development and significant success of DFT are associated with the development of hybrid functionals, in which a nonlocal HF exchange operator is added with a certain weight to the local potential. These are most widely used B3LYP (3 in the designation means the use of a three-parametric formula) and PBE0 functionals. For example, the B3LYP functional has the form

$$E_{XC} = E_{XC}^{\mathrm{LDA}} + a_0(E_X^{\mathrm{HF}} - E_X^{\mathrm{LDA}}) + a_X \Delta E_X^{\mathrm{B88}} + a_C \Delta E_C^{\mathrm{PW91}}, \qquad (3.221)$$

where Δ means that gradient corrections to LDA are taken.

Here, we also mention another important problem with DFT: it does not describe dispersion interactions as a result of the local character of exchange correlation potentials. Reliable structures and binding energies of molecular structures including van der Waals complexes in the ground electronic state can be obtained using the DFT method with an empirical dispersion correction (DFT-D method, see Refs. [69–71]).

3.7 DFT Methods for Calculations of Excited Electronic States: TDDFT

3.7.1 Runge–Gross Theorem

DFT methods are not directly applicable to excited electronic states, because the main postulates of DFT relate to nondegenerate (!) ground-state wave functions and energies. However, it is known that there is an alternative to CI methods for calculations of excited states in common HF theory. This alternative is presented by the time-dependent Hartree–Fock (TDHF) method, which is used for calculations of excited states in the physics of many-body systems [63]. The TDHF equations are called the random-phase approximation (RPA), while the CIS equations are called the Tamm–Dancoff approximation (TDA). In fact, the application of DFT to excited electronic states is based on an analogue of TDHF equations developed for the case of DFT. This theory is, therefore, called "time-dependent density functional theory."

TDDFT is based on the Runge–Gross theorem [64], stating that the exact time-dependent electron density function $\rho(r,t)$ determines the time-dependent external potential $V(r,t)$ up to a time-dependent constant $C(t)$ and, thus, determines the time-dependent wave function up to a phase factor

$$\Psi(r,t) = \Psi[\rho(t)](t)e^{-i\alpha(t)},\ (d/dt)\alpha(t) = C(t). \tag{3.222}$$

The proof of the Runge–Gross theorem is given in their original paper [72] (see also the review in Ref. [34]).

3.7.2 Time-Dependent Kohn–Sham Equations

Let us present time-dependent one-electron wave functions in the basis set of time-independent one-electron functions $\{\chi_i(r)\}$ so that

$$\phi_p(r,t) = \sum_i^M c_{pj}\chi_j(r). \tag{3.223}$$

Then, time-dependent KS equations can be written in matrix designations as follows:

$$i\frac{\partial}{\partial t}\mathbf{C} = \mathbf{F}^{\mathrm{KS}}\mathbf{C} \tag{3.224}$$

Here, the column of matrix C with number i contains time-dependent expansion coefficients $\phi_i(r,t)$, and $\mathbf{F}^{\mathrm{KS}}$ is a matrix representation of the time-dependent KS operator (see Ref. [34]).

Let us suggest that an alternating electric field acts on the system whose matrix elements can be written as follows:

$$g_{pq} = \frac{1}{2}\left[f_{pq}e^{-i\omega t} + f_{qp}^{*}e^{i\omega t}\right] \tag{3.225}$$

The equations are derived using the linear response approximation, that is, within the first-order perturbation theory with respect to the field. After long but simple transformations, the following matrix equations are obtained:

$$\begin{bmatrix} \mathbf{A} & \mathbf{B} \\ \mathbf{B}^* & \mathbf{A}^* \end{bmatrix}\begin{bmatrix} \mathbf{X} \\ \mathbf{Y} \end{bmatrix} = \omega\begin{bmatrix} 1 & 0 \\ 0 & -1 \end{bmatrix}\begin{bmatrix} \mathbf{X} \\ \mathbf{Y} \end{bmatrix} \tag{3.226}$$

Matrix elements of **A** and **B** are defined as

$$A_{ia,jb} = \delta_{ij}\delta_{ab}(\varepsilon_a - \varepsilon_i) + (ia\,|\,jb) + (ia\,|\,f_{xc}\,|\,jb),$$

$$B_{ia,jb} = (ia\,|\,bj) + (ia\,|\,f_{xc}\,|\,bj). \tag{3.227}$$

Two-electron integrals here are written in the chemical notation.

Note that the same scheme of considerations is applicable to common HF theory with the difference that the **A** and **B** matrices are defined somewhat differently, namely

$$A_{ia,jb} = \delta_{ij}\delta_{ab}(\varepsilon_a - \varepsilon_i) + (ia \mid jb),$$

$$B_{ia,jb} = (ia \mid bj). \tag{3.228}$$

If we neglect matrix **B** in the TDDFT equations, we obtain Hermitian equations similar to TDA (i.e., CIS)

$$\mathbf{AX} = \omega\mathbf{X}. \tag{3.229}$$

If matrix **A** – **B** is positively defined, the TDDFT equations can also be transformed to a Hermitian form

$$(\mathbf{A} - \mathbf{B})^{1/2}(\mathbf{A} + \mathbf{B})(\mathbf{A} - \mathbf{B})^{1/2}\mathbf{Z} = \omega^2\mathbf{Z}, \tag{3.230}$$

where

$$\mathbf{Z} = (\mathbf{A} - \mathbf{B})^{1/2}(\mathbf{X} + \mathbf{Y}). \tag{3.231}$$

TDDFT turned out to be rather successful in calculations of electronic spectra of a great number of compounds, including various dyes. However, problems arose in the consideration of systems in which electronic transitions are characterized by charge transfer (CT).

As opposite to the case of CIS, energies of CT transitions turn out to be strongly underestimated, which directly follows from the structure of matrix elements of matrices **A**. This problem is thoroughly discussed in Ref. [34] (see also Ref. [65]).
Actually,

$$A_{ia,ia} = (\varepsilon_a - \varepsilon_i) + (ia \mid ia)). \tag{3.232}$$

In the case of CT transitions, when orbitals ψ_i and ψ_a are spatially separated (one is localized on molecule or fragment *A* and another one on molecule or fragment *B*), they do not overlap, their exchange integral is nearly equal to zero, and, hence, the transition energy equals the difference in the orbital energies of electron-accepting and electron-donating MOs $\varepsilon_a - \varepsilon_i$. In HF theory, this expression gives a rough estimate for the energy of the CT state at large distances, because occupied orbitals are calculated in the field of $(N - 1)$ electrons, while virtual orbitals are calculated in the field of *N* electrons. Therefore, according to Koopmans's theorem, $-\varepsilon_i$ and $-\varepsilon_a$ correspond to the IP of molecule *A* and to the electron affinity (EA) of molecule *B*, respectively.

The situation is different in the KS DFT formalism, in which the same (local!) potential is used for both occupied and virtual orbitals. Consequently, while the HOMO energy in KS DFT corresponds to the IP as well as in HF theory, the lowest unoccupied molecular orbital (LUMO) energy is determined in the same field as the HOMO energy, and this field is specially parametrized so as to take into account the interaction of an electron with remaining (N – 1) electrons. Therefore, the LUMO is generally more strongly bound in DFT that in HF theory and cannot be related to the EA, the negative of the LUMO energy is much larger than the true EA calculated as difference in the total energies of the charged and neutral systems, and the orbital energy difference corresponding to a CT state is commonly a drastic underestimation of the correct excitation or CT energy. Moreover, because the energy of excitation to a CT state simply equals the constant difference of the corresponding orbital energies $\varepsilon_a - \varepsilon_i$, the potential energy curves of CT states do not exhibit the correct $1/R$ behavior but are constant.

Because the main problems in the DFT description of excited states and excitation energies are associated with the locality of the KS DFT exchange functional, many DFT functionals were suggested that include nonlocal (HF) exchange into the DFT exchange correlation potential. Very often, the extent of inclusion of nonlocal exchange depends on the distance R (so-called long-range corrected functionals [66, 67]). As a matter of fact many well-parametrized DFT functionals (e.g., B3LYP) provide rather good equilibrium structures of molecules and their thermochemistry.

3.8 Valence Approximation and Pseudopotentials

3.8.1 Separation of Core and Valence Electrons: Frozen-Core Approximation

We have already considered the problem of separation of variables in molecular quantum mechanics in the case of adiabatic approximation. Now, we will consider the situation when an electronic system can be divided into (two) subsystems. The most important case is the division of the electronic system of a molecule into the core and valence subsystems. This division is somewhat arbitrary and depends on the elements constituting the molecule. Commonly, AOs that essentially participate in the formation of chemical bonds are included into the valence shell. All lower-lying AOs are included into the core shell.

We may suggest that the total molecular electronic wave function can be written as an antisymmetrized product of wave functions of the two subsystems

$$\Psi(1,2,...,n) = A\Phi_{\text{core}}(1,2,...,n_{\text{c}})\Phi_{\text{val}}(1,2,...,n_{\text{v}}). \qquad (3.233)$$

In the case of an HF wave function, this division is exact. However, in a general case, this assumption can be considered as a physically justified approximation, if the energy levels of core electrons are considerably lower than the energy levels of valence electrons. Another important case is a spatial separation, when correlations of electrons from different subsystems are insignificant.

If the core wave function $F_{\text{core}}(1,2,...,n_{\text{c}})$ is known, we can write an equation for the valence subsystem

$$\hat{H}_{\text{val}}\Phi_{\text{val}} = E_{\text{val}}\Phi_{\text{val}}. \qquad (3.234)$$

Here, $E_{\text{val}} = E - E_{\text{core}}$, and

$$\hat{H}_{\text{val}} = \sum_{i=1}^{n_{\text{val}}}\left(h_i + \sum_{j\neq i} r_{ij}^{-1}\right) + \hat{V}_{\text{core}}, \qquad (3.235)$$

where $\hat{V}_{\text{core}}$ is an operator describing the interaction of core and valence electrons. It is, generally, a nonlocal integral operator, because the occurrence of the antisymmetrizer in Eq. 3.233 gives rise to exchange terms.

The valence wave function must be orthogonal to the core one. In the case of the HF approximation, the orthogonality of the wave functions is ensured by the orthogonality of the orbitals of the corresponding subsystems. It is possible to derive equations for HF orbitals in a more general case, when we do not require the orthogonality between the valence and core orbitals in an explicit form. Let us consider a system with a closed shell and divide the entire set of doubly occupied MO into two, core $\{\phi_{\text{c}}^i\}$ and valence $\{\phi_{\text{v}}^i\}$ subsystems.

In the frozen-core approximation, the set $\left\{\phi_{\mathrm{c}}^{i}\right\}$ determines the core subspace R_{c}, which is assumed to be fixed. The valence MOs belong to the orthogonal complement of R_{c}, namely to the valence subspace R_{v}. The total energy functional must be minimized with respect to any variations of $\left\{\phi_{\mathrm{v}}^{i}\right\}$ in the valence subspace. The required one-electron equations for valence MOs are simply obtained by projecting the HF equations onto the valence subspace

$$\hat{P}_{\mathrm{v}}\hat{F}\hat{P}_{\mathrm{v}}\phi_{\mathrm{v}}^{i} = \varepsilon_{\mathrm{v}}^{i}\phi_{\mathrm{v}}^{i}. \tag{3.236}$$

Here, $\hat{P}_{\mathrm{v}} = 1 - \hat{P}_{\mathrm{c}} = 1 - \sum_{i}\left|\phi_{\mathrm{c}}^{i}\right\rangle\left\langle\phi_{\mathrm{c}}^{i}\right|$ is the projector onto the R_{v} subspace, and $\hat{F}$ is the Fockian defined in Eq. 3.131.

3.8.2 Pseudovalence MO, Phillips–Kleinman Equations, and Effective Core Potential

Now, we will introduce pseudovalence orbitals ψ_{v}^{i} such that

$$\phi_{\mathrm{v}}^{i} = \hat{P}_{\mathrm{v}}\psi_{\mathrm{v}}^{i}. \tag{3.237}$$

Then, we will obtain equations for pseudovalence MOs

$$\hat{P}_{\mathrm{v}}\hat{F}\hat{P}_{\mathrm{v}}\psi_{\mathrm{v}}^{i} = \varepsilon_{\mathrm{v}}^{i}P_{\mathrm{v}}\psi_{\mathrm{v}}^{i}. \tag{3.238}$$

Here, we used the main property of the projection operator $\hat{P}^{2} = \hat{P}$. Now, we can obtain the generalized Phillips–Kleinman (GPK) equations [68]. For this purpose, it is sufficient to use the equality

$$\hat{P}_{\mathrm{v}} = 1 - \hat{P}_{\mathrm{c}} \tag{3.239}$$

and substitute it into Eq. 3.238

$$\left(1 - \hat{P}_{\mathrm{c}}\right)\hat{F}\left(1 - \hat{P}_{\mathrm{c}}\right)\psi_{\mathrm{v}}^{i} = \varepsilon_{\mathrm{v}}^{i}\left(1 - \hat{P}_{\mathrm{c}}\right)\psi_{\mathrm{v}}^{i}. \tag{3.240}$$

If we now remove the parentheses, we will obtain the following equations:

$$\left(\hat{F} + V_{i}^{\mathrm{GPK}}\right)\psi_{\mathrm{v}}^{i} = \varepsilon_{\mathrm{v}}^{i}\psi_{\mathrm{v}}^{i}, \tag{3.241a}$$

where

$$V_{i}^{\mathrm{GPK}} = -\hat{P}_{\mathrm{c}}\hat{F} - \hat{F}\hat{P}_{\mathrm{c}} + \hat{P}_{\mathrm{c}}\hat{F}\hat{P}_{\mathrm{c}} + \varepsilon_{\mathrm{v}}^{i}\hat{P}_{\mathrm{c}}. \tag{3.241b}$$

Equations 3.241a and 3.241b define the GPK pseudopotential V_i^{GPK}. We have not suggested so far that the core functions are solutions of the HF equations. If, however, such is the case, then

$$\hat{F}\hat{P}_{\mathrm{c}} = \hat{P}_{\mathrm{c}}\hat{F} = \sum_i \varepsilon_{\mathrm{c}}^i \left|\phi_{\mathrm{c}}^i\right\rangle\left\langle\phi_{\mathrm{c}}^i\right|, \tag{3.242}$$

and we obtain

$$V_i^{\mathrm{GPK}} = -\sum_j \left(\varepsilon_{\mathrm{c}}^j - \varepsilon_{\mathrm{v}}^i\right)\left|\phi_{\mathrm{c}}^j\right\rangle\left\langle\phi_{\mathrm{c}}^j\right|, \tag{3.243}$$

which is the common form of the GPK or, simply, the Phillips–Kleinman (PK) pseudopotential. Remember here that the HF operator can be presented in the form (see Eq. 3.131)

$$\hat{F} = \hat{F}_{\mathrm{v}} + \left(2\hat{J}_{\mathrm{c}} - \hat{K}_{\mathrm{c}}\right), \tag{3.244}$$

so that we obtain

$$\left[\hat{F}_{\mathrm{v}} + \left(2\hat{J}_{\mathrm{c}} - \hat{K}_{\mathrm{c}}\right) + V_i^{\mathrm{PK}}\right]\psi_{\mathrm{v}}^i = \varepsilon_{\mathrm{v}}^i \psi_{\mathrm{v}}^i. \tag{3.245}$$

We can introduce the effective core potential (ECP)

$$V^{\mathrm{ECP}} = \left(2\hat{J}_{\mathrm{c}} - \hat{K}_{\mathrm{c}}\right) + V_i^{\mathrm{PK}}, \tag{3.246}$$

and, finally, obtain equations for pseudovalence orbitals in the form

$$\left[\hat{F}_{\mathrm{v}} + V^{\mathrm{ECP}}\right]\psi_{\mathrm{v}}^i = \varepsilon_{\mathrm{v}}^i \psi_{\mathrm{v}}^i. \tag{3.247}$$

We note here that, first, the PK pseudopotential depends on the energy of the valence MO on which it acts, and, second, it is a nonlocal operator. Moreover, the solution of the PK equations is ambiguous by their construction, because it is determined up to any linear combination of core functions. Therefore, the PK pseudopotential is not used in real molecular calculations. However, it still remains a theoretical basis for the construction of various model ECPs.

3.8.3 Practical Implementation of Pseudo- and Model Potentials

With the exception of rather primitive local approximations, one of the first widely used approximations were model Huzinaga-type potentials of the form

$$\hat{V}_i^{\mathrm{H}} = -\frac{\left(Z - n_{\mathrm{c}}\right)}{r_i} - \sum_j A_j \frac{e^{-\alpha_j r_i^2}}{r_i} + \sum_c B_{\mathrm{c}}\left|\phi_{\mathrm{c}}\right\rangle\left\langle\phi_{\mathrm{c}}\right|. \tag{3.248}$$

(see Refs. [69, 70]). Kahn and Goddard proposed model potentials in the form of projectors onto subspaces of atomic functions with a given angular quantum number in the form [71, 72]

$$\left(-\tfrac{1}{2}\nabla^2 - \tfrac{Z}{r} + \frac{l(l+1)}{2r^2} + V_{\text{val}} + V_{\text{core}}\right)\phi_l = \varepsilon_l \phi_l . \qquad (3.249)$$

or

$$\left(-\tfrac{1}{2}\nabla^2 - \frac{Z^{\text{eff}}}{r} + \frac{l(l+1)}{2r^2} + V_{\text{val}}^{'} + V_l^{\text{eff}}\right)\chi_l = \varepsilon_l \chi_l . \qquad (3.250)$$

Here, χ_l is a pseudoorbital constructed from HF orbitals, and V_l^{eff} is the ECP. The stroke at V_{val} means that the valence potential is determined from the valence pseudoorbitals rather than the initial valence orbitals. If χ_l are nodeless functions, Equation 3.250 can be inverted, so V_l^{eff} can be directly found from this equation. Model potentials of the Kahn–Goddard type turned out to be more flexible and universal.

The total ECP can be written in the form of a sum of projectors onto the corresponding angular subspaces

$$V^{\text{eff}} = \sum_{l=0}^{l_{\max}} V_l^{\text{eff}} \sum_m |lm\rangle\langle lm| . \qquad (3.251)$$

The potentials for orbitals with $l > l_{\max}$ are virtually identical, so we may assume that

$$V^{\text{eff}} = V_{l_{\max}}^{\text{eff}} + \sum_{l=0}^{l_{\max}-1} \left(V_l^{\text{eff}} - V_{l_{\max}}^{\text{eff}}\right) \sum_m |lm\rangle\langle lm| , \qquad (3.252)$$

where $l_{\max}$ is by unity larger than the largest value of l in the core. Then, the problem is reduced to the selection of nodeless pseudovalence functions χ_l.

Let us note in conclusion of this section that only electrons residing in the outermost (valence) shells of heavy atoms are responsible for most properties of chemical interest. The atomic inner core shells remain to a large extent unchanged when molecules are formed from atoms. Moreover, relativistic effects, not only those of the core electrons, but also those directly associated with valence electrons, arise in regions near the nuclei where the local kinetic energy of the electrons is high. Assuming that in both cases the relevant information can be transferred from atoms to molecules, one is led to considering a model system of n_v formally nonrelativistic valence electrons interacting with fixed cores, described by a Hamiltonian like

$$H_{\mathrm{PP}} = \sum_{i}^{n_v} \left[-\tfrac{1}{2}\vec{p}_i^{\,2} + \sum_{\lambda} V_{\lambda}^{\mathrm{PP}}(r_{i\lambda}) \right] + \sum_{i<j}^{n_v} \frac{1}{r_{ij}} + \sum_{\lambda<\mu} \frac{Q_\lambda Q_\mu}{r_{\lambda\mu}}. \quad (3.253)$$

For further reading on pseudopotentials and model potentials, refer to some review articles [73–76].

References

1. V. Magnasco, *Elementary Molecular Quantum Mechanics: Mathematical Methods and Applications*, 2nd ed., Elsevier, Amsterdam, 2013, 932 p.
2. H. Eyring, J. Walter, and G. E. Kimball, *Quantum Chemistry*, John Wiley and Sons, New York, 1947.
3. A. Szabo and N. S. Ostlund, *Modern Quantum Chemistry: Introduction to Advanced Electronic Structure Theory*, McGraw-Hill, New York, 1989.
4. J. Simons and J. Nichols, *Quantum Mechanics in Chemistry, Topics in Physical Chemistry*, Oxford University Press, New York, 1997.
5. R. McWeeny, *Methods of Molecular Quantum Mechanics*, 2nd ed., Academic Press, London, 1992.
6. D. A. McQuarrie, *Quantum Chemistry*, 2nd ed., University Science Book, Sauisalito, 2008.
7. W. Pauli, Jr., in *Handbuch der Physik*, Springer-Verlag, Berlin, 1933, Paragraph 11.
8. J. I. Musher, Comment on some theorems of quantum chemistry, *Am. J. Phys.*, 1966, **34**, 267.
9. R. P. Feynman, Forces in molecules, *Phys. Rev.*, 1939, **56**, 340.
10. E. P. Wigner, On a modification of the Rayleigh-Schrödinger perturbation theory, *Math. Natur. Anz. (Budapest)*, 1935, **53**, 477–482.
11. M. Born and R. Oppenheimer, Zur quantentheorie der molekeln, *Ann. Phys.*, 1927, **84**, 457.
12. B. Valeur, *Molecular Fluorescence: Principles and Applications*, Wiley-VCH, Weinheim, 2002.
13. C. M. Marian, Spin–orbit coupling and intersystem crossing in molecules, *WIREs Comput. Mol. Sci.*, 2012, **2**, 187–203.

14. C. M. Marian, Spin–orbit coupling in molecules, in *Reviews in Computational Chemistry*, Vol. 17, K. B. Lipkowitz and D. B. Boyd (eds.), Wiley-VCH, New York, 2001, pp. 99–204.

15. D. G. Fedorov, S. Koseki, M. W. Schmidt, and M. S. Gordon, Spin-orbit coupling in molecules: chemistry beyond the adiabatic approximation, *Int. Rev. Phys. Chem.*, 2003, **22**(3), 551–592.

16. R. S. Mulliken, *Life of a Scientist*, B. J. Ransil (ed.), Springer, Berlin, Heidelberg, 1989, 256 p.

17. http://winter.group.shef.ac.uk/orbitron/

18. D. R. Hartree, The wave mechanics of an atom with a non-Coulomb central field. Part I. Theory and methods, *Proc. Cambridge Philos. Soc.*, 1928, **24**, 89.

19. P. O. Löwdin, Quantum theory of many-particles systems, I: physical interpretations by mean of density matrices, natural spin-orbitals, and convergence problems in the method of configurational interaction, *Phys. Rev.*, 1955, **97**, 1474–1489.

20. I. Mayer, *Simple Theorems, Proofs, and Derivations in Quantum Chemistry*, Springer, New York, 337 p.

21. V. Fock, Näherungsmethode zur Lösung des quantenmechanischen Mehrkörperproblems, *Z. Phys.*, 1930, **61**, 126.

22. C. C. J. Roothaan, New developments in molecular orbital theory, *Rev. Mod. Phys.*, 1951, **23**, 69–89.

23. C. C. J. Roothaan, Self-consistent field theory for open shells of electronic systems, *Rev. Mod. Phys.*, 1960, **32**, 179–185.

24. S. F. Boys, Electronic wave functions. I. A general method of calculation for the stationary states of any molecular system, *Proc. R. Soc. Lond. A*, 1950, **200**, 542–554.

25. T. Koopmans, Über die Zuordnung von Wellenfunktionen und Eigenwerten zu den Einzelnen Elektronen eines Atoms, *Physica*, 1934, **1**(1–6), 104–113.

26. S. Huzinaga, Applicability of Roothaan's self-consistent field theory, *Phys. Rev.*, 1960, **120**(3), 866–871.

27. S. Huzinaga, Analytical methods in Hartree-Fock self-consistent field theory, *Phys. Rev.*, 1961, **122**(3), 131–138.

28. B. N. Plakhutin, Symmetrical conditions for the state coupling coefficients in the open-shell SCF theory, in *Reviews of Modern Quantum Chemistry: A Celebration of the Contributions of Robert G. Parr*, Vol. 1 (in 2 volumes), K. D. Sen (ed.), World Scientific, New Jersey, London, Singapore, 2002, pp. 16–42.

29. J. L. Dodds and R. McWeeny, Orbital energies and Koopmans' theorem in open shell Hartree-Fock theory, *Chem. Phys. Lett.*, 1972, **13**(1), 9–12.
30. B. N. Plakhutin and E. R. Davidson, Koopmans' theorem in the restricted open-shell Hartree-Fock method. 1. A variational approach, *J. Phys. Chem. A*, 2009, **113**, 12386–12395.
31. E. R. Davidson and B. N. Plakhutin, Koopmans's theorem in the restricted open-shell Hartree–Fock method. II. The second canonical set for orbitals and orbital energies, *J. Chem. Phys.*, 2010, **132**, 184110.
32. C. D. Sherrill and H. F. Schaefer III, The configuration interaction method: Advances in highly correlated approaches, *Adv. Quantum Chem.*, 1999, **34**, 143–269.
33. J. B. Foreman, M. Head-Gordon, J. A. Pople, and M. J. Frisch, Toward a systematic molecular orbital theory for excited states, *J. Phys. Chem.*, 1992, **96**, 135–149.
34. A. Dreuw and M. Head-Gordon, Single-reference ab initio methods for the calculation of excited states of large molecules, *Chem. Rev.*, 2005, **105**, 4009–4037.
35. A. Broo and A. Holmén, Calculations and characterization of the electronic spectra of DNA bases based on ab initio MP2 geometries of different tautomeric forms, *J. Phys. Chem. A*, 1997, **101**, 3589–3600.
36. M. K. Shukla, S. K. Mishra, A. Kumar, and P. C. Mishra, An ab initio study of excited states of guanine in the gas phase and aqueous media: electronic transitions and mechanism of spectral oscillations, *J. Comp. Chem.*, 2000, **21**, 826–846.
37. J. E. Subotnik, Communication: configuration interaction singles has a large systematic bias against charge-transfer states, *J. Chem. Phys.*, 2011, **135**, 071104.
38. P. E. M. Siegbahn, J. Almlof, A. Heiberg, and B. O. Roos, The complete active space SCF (CASSCF) method in a Newton–Raphson formulation with application to the HNO molecule, *J. Chem. Phys.*, 1981, **74**(4), 2384–2396.
39. J. Olsen, The CASSCF method: a perspective and commentary, *Int. J. Quantum Chem.*, 2011, **111**, 3267–3272.
40. C. Møller and M. S. Plesset, Note on an approximation treatment for many-electron systems, *Phys. Rev.*, 1934, **46**(7), 618–622.
41. H.G. Kümmel, A biography of the coupled cluster method, *Int. J. Mod. Phys. B*, 2003, **17**, 5311–5325.
42. J. Hubbard, The description of collective motions in terms of many-body perturbation theory, *Proc. R. Soc. Lond. A*, 1957, **240**, 539.

43. A. I. Krylov, Equation-of-motion coupled-cluster methods for open-shell and electronically excited species: the Hitchhiker's guide to Fock space, *Annu. Rev. Phys. Chem.*, 2008, **59**, 433–462.

44. M. Head-Gordon, R. J. Rico, M. Oumi, and T. J. Lee, A doubles correction to electronic excited states from configuration interaction in the space of single substitutions, *Chem. Phys. Lett.*, 1994, **219**, 21–29.

45. M. Head-Gordon, D. Maurice, and M. Oumi, A perturbative correction to restricted open shell configuration interaction with single substitutions for excited states of radicals, *Chem. Phys. Lett.*, 1995, **246**, 114–121.

46. R. G. Parr and W. Yang, *Density-Functional Theory of Atoms and Molecules*, Oxford University Press, New York, 1989.

47. N. H. March, *Electron Density Theory of Atoms and Molecules*, Academic Press, New York, 1992.

48. P. Hohenberg and W. Kohn, Inhomogeneous electron gas, *Phys. Rev.*, 1964, **136**, B864–B871.

49. W. Kohn, V-representability and density functional theory, *Phys. Rev. Lett.*, 1983, **51**(17), 1596.

50. W. Kohn and L. J. Sham, Self-consistent equations including exchange and correlation effects, *Phys. Rev. A*, 1965, **140**, 1133–1138.

51. M. Levy, J. P. Perdew, and V. Sahni, Exact differential equation for the density and ionization energy of a many-particle system, *Phys. Rev. A*, 1984, **30**, 2745.

52. J. P. Perdew and M. Levy, Physical content of the exact Kohn-Sham orbital energies: band gaps and derivative discontinuities, *Phys. Rev. Lett.*, 1983, **51**(20), 1884–1887.

53. T. Tsuneda, *Density Functional Theory in Quantum Chemistry*, Springer, Tokyo, 2014 200 p.

54. E. Fermi, Eine statistische Methode zur Bestimmung einiger Eigenschaften des Atoms und ihre Anwendung auf die Theorie des periodischen Systems der Elemente, *Z. Phys.*, 1928, **48**, 73–79.

55. D. M. Ceperley and B. J. Alder, Ground state of the electron gas by a stochastic method, *Phys. Rev. Lett.*, 1980, **45**, 566–569.

56. S. J. Vosko, L. Wilk, and M. Nusair, Accurate spin-dependent electron liquid correlation energies for local spin density calculations: a critical analysis, *Can. J. Phys.*, 1980, **58**, 1200.

57. F. Jensen, Introduction *to* Computational Chemistry, Wiley, Chichester, 1999, 429 p.

58. R. G. Parr and W. Yang, *Density-Functional Theory of Atoms and Molecules*, Oxford University Press, New York, 1989, 333 p.

59. B. Y. Tong and L. J. Sham, Application of a self-consistent scheme including exchange and correlation effects to atoms, *Phys. Rev.*, 1966, **144**(1), 1.

60. J. P. Perdew, J. A. Chevary, S. H. Vosko, K. A. Jackson, M. R. Pederson, D. J. Singh, and C. Fiolhais, Atoms, molecules, solids, and surfaces: applications of the generalized gradient approximation for exchange and correlation, Phys. Rev. B, 1992, 46(11), 6671.

61. A. D. Becke, Density-functional thermochemistry. III. The role of exact exchange, *J. Chem. Phys.*, 1993, **98**(7), 5648.

62. D. C. Langreth and M. J. Mehl, Beyond the local-density approximation in calculations of ground-state electronic properties, *Phys. Rev. B*, 1983, **28**(4), 1809.

63. J. P. Perdew and Y. Wang, Accurate and simple density functional for the electronic exchange energy: generalized gradient approximation, *Phys. Rev. B*, 1986, **33**, 8800–8802.

64. J. P. Perdew, K. Burke, and M. Ernzerhof, Generalized gradient approximation made simple, *Phys. Rev. Lett.*, 1996, **77**(18), 3865–3868.

65. A. D. Becke, Density-functional exchange-energy approximation with correct asymptotic behavior, *Phys. Rev. A*, 1988, **38**(6), 3098–3100.

66. R. Colle and O. Salvetti, Approximate calculation of the correlation energy for the closed shells, *Theoret. Chim. Acta (Berl.)*, 1975, **37**, 329–334.

67. C. Lee, W. Yang, and R. G. Parr, Development of the Colle-Salvetti correlation-energy formula into a functional of the electron density, *Phys. Rev. B*, 1988, **37**(2), 785–789.

68. S. Grimme, Accurate description of van der Waals complexes by density functional theory including empirical corrections, *J. Comput. Chem.*, 2004, **25**, 1463–1473.

69. S. Grimme, Semiempirical GGA-type density functional constructed with a long-range dispersion correction, *J. Comput. Chem.*, 2006, **27**, 1787–1799.

70. S. Grimme, J. Antony, S. Ehrlich, and H. Krieg, A consistent and accurate ab initio parametrization of density functional dispersion correction (DFT-D) for the 94 elements H-Pu, *J. Chem. Phys.*, 2010, **132**, 154104.

71. D. J. Thouless, *The Quantum Mechanics of Many-Body Systems*, Academic Press, New York, London, 1961, 175 p.

72. E. Runge and E. K. U. Gross, density-functional theory for time-dependent systems, *Phys. Rev. Lett.*, 1984, **52**(12), 997–1000.

73. C. A. Ullrich, *Time-Dependent Density-Functional Theory: Concepts and Applications*, Oxford University Press, New York, 2012, 544 p.
74. H. Iikura, T. Tsuneda, T. Yanai, and K. Hirao, A long-range correction scheme for generalized-gradient approximation exchange functionals, *J. Chem. Phys.*, 2001, **115**, 3540–3544.
75. Y. Tawada, T. Tsuneda, S. Yanagisawa, T. Yanai, and K. Hirao, A long-range-corrected time-dependent density functional theory, *J. Chem. Phys.*, 2004, **120**, 8425–8433.
76. J. C. Phillips and L. Kleinman, New method for calculating wave functions in crystals and molecules, *Phys. Rev.*, 1959, **116**, 287.
77. V. Bonifacic and S. Huzinaga, Atomic and molecular calculations with the model potential method. I, *J. Chem. Phys.*, 1974, **60**, 2779–2786.
78. V. Bonifacic and S. Huzinaga, Atomic and molecular calculations with the model potential method. II, *J. Chem. Phys.*, 1975, **62**, 1507–1508.
79. L. R. Kahn and W. A. Goddard III, A direct test of the validity of the use of pseudopotentials in molecules, *Chem. Phys. Lett.*, 1968, **2**, 667.
80. L. R. Kahn and W. A. Goddard III, Ab initio effective potentials for use in molecular calculations, *J. Chem. Phys.*, 1972, **56**, 2685–2701.
81. K. G. Dyall, Formal analysis of effective core potential methods, *J. Chem. Inf. Comput. Sci.*, 2001, **41**, 30–37.
82. M. Krauss and W. J. Stevens, Effective potentials in molecular quantum chemistry, *Ann. Rev. Phys. Chem.*, 1984, **35**, 357–385.
83. M. Dolg and X. Cao, Relativistic pseudopotentials: their development and scope of applications, *Chem. Rev.*, 2012, **112**, 403–480.
84. P. Schwerdtfeger, The pseudopotential approximation in electronic structure theory, *ChemPhysChem*, 2011, **12**, 3143–3155.

Chapter 4

Classical Methods of Molecular Simulations of Disordered Materials

4.1 Introduction

Fast development of computational hardware and corresponding software made possible computer simulation of micro- and macroscopic properties of various condensed systems from classical nonideal liquids to solvated complexes of biological macromolecules. Physical interpretation of results of computer simulation suggests a knowledge of theoretical basics of computational methods of statistical physics and practical experience in the use of molecular dynamics (MD) and Monte Carlo (MC) methods in calculations of the structure and properties of disordered condensed systems. As a rule, the available books and tutorials present in detail either theoretical basics of statistical physics or implementation procedures for a particular method of computer simulation, like MD or MC. In this chapter, an attempt is made to expound elements of statistical mechanics in connection with methodology of MD simulation in order that the obtained knowledge can be applied in practical calculations of disordered condensed systems using Groningen Simulation Package (GROMACS) or related software. Special attention is paid to theories and approaches that describe the interconnection of macroscopic characteristics of a substance with its molecular structure.

Multiscale Modeling in Nanophotonics: Materials and Simulation
Alexander Bagaturyants and Mikhail Vener

ISBN 978-981-4774-40-6 (Hardcover), 978-1-315-10976-3 (eBook)
www.panstanford.com

4.2 Elements of Statistical Mechanics

4.2.1 Gnoseology of Microscopic Objects

The general character of laws that govern properties of a macroscopic object is independent of the mechanics that describe the motion of its constituent particles [1]. This statement justifies the use of classical mechanics in computational methods of statistical physics.

For conservative systems, the Hamiltonian function equals a sum of the kinetic and potential energies, that is, equals the total energy of the system in variables q and p[a]

$$H(q,p) = K(q,p) + U(q) = E\,. \tag{4.1}$$

Here, q are generalized coordinates and p are generalized momenta. For a particle moving in a stationary field $U = U(x,y,z)$ $H = \frac{1}{2m}\left(p_x^2 + p_y^2 + p_z^2\right) + U(x,y,z)$, the Hamiltonian equations take the form[b]

$$\frac{\partial H}{\partial p_i} = \dot{q}_i\,, \frac{\partial H}{\partial q_i} = -\dot{p}_i\,. \tag{4.2}$$

In the Hamiltonian approach, canonical variables q_i and p_i are independent variables; therefore, the number of equations in Eq. 4.1 equals $6N$, where N is the number of particles, while the Hamiltonian equations are first-order equations. A conservative closed systems with $3N$ degrees of freedom has $6N - 1$ integrals of motion. In statistical physics, a special role is played by the energy

$$H(p,q) = E = \text{const.} \tag{4.3}$$

The motion of mechanical systems is described using the concept of phase space (Γ space). For a system with $3N$ degrees of freedom, the Γ space is an abstract space of $6N$ dimensions ($3N$ generalized coordinates and $3N$ generalized momenta). The current state of the system is described by a point in the Γ space (representation or phase point). Then, the motion of the system is described by a 1D curve in the $6N$-dimensional Γ space (phase trajectory), whose parametric equation has the form

---

[a]For more details, see Chapter 2; for the convenience of the readers, we give here a brief account of this topic.

[b]See Chapter 2, Eq. 2.55.

$$qi = q_i(t;q^0,p^0),\ p_i = p_i\ (t;q^0,p^0),\ i = 1,2,\ldots,3N, \tag{4.4}$$

where q^0 and p^0 are the initial values of coordinates and momenta.

A closed and conservative system has the energy integral shown in Eq. 4.3. In this case, the 1D phase trajectory lies on a $(6N - 1)$-dimensional hypersurface, whose equation has the form of Eq. 4.3. The elementary volume of the Γ space whose points lie between q_i, p_i and $q_i + dq_i$, $p_i + dp_i$, that is, inside an infinitesimally small $6N$-dimensional parallelepiped with edges dq_i, dp_i equals

$$d\Gamma = dq_1 dq_2 ... dq_s dp_1 dp_2 ... dp_s = \prod_{i=1}^{s} dq_i dp_i = (dq)(dp)\,. \tag{4.5}$$

From here, it follows that the dimensionality of the Γ space for a system with s degrees of freedom equals $[\text{action}]^s$, where the dimensionality of action is [action] = L^2MT^{-1}.

In some cases (ideal gas), it is convenient to introduce a phase space corresponding to coordinates and conjugate momenta of a *single* molecule. For a monoatomic gas, the position of whose atoms is determined by Cartesian (orthogonal) coordinates x_i, y_i, z_i, this is a 6D space with axes x, y, z, p_x, p_y, p_z. The state of an ideal gas of N molecules is represented by N phase points. A uniformly moving gas atom is described in the 6D phase space by a trajectory in the form of a straight line perpendicular to p_x, p_y, and, p_z axes.

To justify methods of statistical physics, it seems essential to consider the motion of a set of a great number N of identical (i.e., with the same $H(p,q)$) systems. This set, named *ensemble*, is described by N phase points in the Γ space. For large values of N, the notion of *distribution function* is introduced

$$\rho(p,q,t) = \frac{\Delta N}{\Delta\Gamma}, \tag{4.6}$$

where ΔN is the number of phase points in a small volume $\Delta\Gamma$ of the phase space. The distribution function is normalized so that

$$\int \rho(p,q,t)\Delta\Gamma = \int \Delta N = N\,. \tag{4.7}$$

Phase trajectories of representation points of an ensemble cannot intersect or be tangent to each other. Otherwise, the mechanical system being in a certain state could move after the intersection or tangency point in different ways, which contradicts to the principle of mechanical causality (i.e., the uniqueness of the solution of the

Hamiltonian equations). The motion of phase points obeys Liouville's theorem. The essence of this theorem is in the following equivalent statements:

- For a given ensemble, points do not disappear and do not arise.
- Representation points move in the phase space as an incompressible liquid; that is, $\rho(q,p,t)$ is constant along the phase trajectory.
- The phase volume $\Delta\Gamma$ containing a given set of moving systems of an ensemble is constant.

Each thermodynamic system occurring in specified external conditions eventually reach a state of thermodynamic (statistical) equilibrium. The volume V, temperature T, and pressure P (for nonisolated systems) are commonly considered as external conditions (parameters). For nonequilibrium systems, gradients of forces (potentials) acting on particles of the system also serve as external parameters.

Along with external parameters, it is also necessary to define internal parameters, the quantities that characterize properties of the system itself. They depend on the external parameters and the temperature (energy) of the system. From a molecular point of view, the internal parameters of the system are defined as functions of the coordinates and momenta of system particles averaged in a corresponding way. In the state of thermodynamic equilibrium of the system, its internal parameters at specified external parameters take on definite numerical values.

The goal of statistical mechanics is to calculate macroscopic quantities, for example, thermodynamic functions, in terms of MD variables. Any change of physical quantity B cannot be performed instantly: it takes a certain time τ. It might be considered that the time average of a dynamic microscopic quantity $b(p,q,t)$ is observed in an experiment

$$B^{\exp} = \langle b \rangle_\tau = \frac{1}{\tau} \int_{t_0}^{t_0+\tau} b(p,q,t)dt\,, \tag{4.8}$$

where τ is the macroscopic observation period. For a system in equilibrium, it might be assumed that $\langle b \rangle_\tau$ does not depend on the initial instant of time τ_0. There is a different formulation: at constant external conditions in an equilibrium system, the experimental value of B found at multiple repetition of the experiment equals a certain average value. Such a computation is connected with the solution of the dynamic problem for the system under consideration, written as the Hamiltonian equations (Eq. 4.2) with the initial conditions $q(0) = q_0$ and $p(0) = p_0$, which are actually unknown for many-particle systems with interaction, that is, for disordered condensed systems. Thus, the solution of the system of differential equations of motion (Eq. 4.3) and, hence, the computation by Eq. 4.8 becomes impossible in principle.

4.2.2 Probabilistic Description of System Microstates: Method of Gibbs Ensembles

The main task of statistical physics is the calculation of the average values of various quantities (pressure, energy, magnetic moment, etc.) for macroscopic bodies in an equilibrium state and also fluctuations of the corresponding values. It is evident that the purely mechanical approach using variables p and q is not sufficient for an analysis of thermodynamic regularities, because fundamental thermodynamic parameters (temperature, entropy, etc.) are not the average values of mechanical quantities.

To disclose the meaning of thermodynamic parameters, it is necessary to pass to a *probabilistic description* of microstates of the system, that is, to consider dynamic variables as random quantities. Then, temperature, entropy, etc., will be interpreted as quantities that characterize the distribution of probabilities of various microstates of the system. The notion of the probability of a certain state of the system is introduced by using the *method of Gibbs ensemble*, that is, by considering an ensemble of systems. An ensemble of systems is a set of a very great number of systems identical in their nature, occurring in similar external conditions, and differing only in their microstate. An ensemble of isolated systems is a set of systems, each of which has prescribed values of N, V, and E: each system is encased in a rigid adiabatic envelop impenetrable for particles, and external force fields are absent. Systems of the ensemble differ only in their mechanical state at a given instant of time (by their phase). The ensemble can be composed by systems exchanging by energy, each of which has given values of N, V, and T.

Statistical physics considers ensembles of such systems interaction among which is absent or negligibly small, so the state of the system and its energy are completely determined by parameters p and q relating to the given system and by external parameters. It is assumed that systems of the ensemble are statistically independent.

It is evident that variables p and q *cannot be determined uniquely* because of indeterminacy in initial conditions. The parameters determining the microstate of ensemble systems are random quantities. This difficulty is overcome in the method of Gibbs ensembles using the following postulate: "Microstates of the system can be provided with definite probabilities (distribution function)." Note that the method of Gibbs ensembles is the only method of studying systems for which the initial microstates are unknown.

The probability $dw(p,q,t)$ that the state of the system at time instant t is characterized by prescribed values of p and q (at given parameters) is given by the equation

$$dw(p,q,t) = \rho(p,q,t)dpdq, \tag{4.9}$$

where $\rho(p,q,t)$ is the probability distribution density in the phase space.

The ensemble average value of some function $b(p,q)$ at time instant t is defined as

$$\overline{b(t)} = \iint b(p,q)\rho(p,q,t)dpdq\,. \tag{4.10}$$

It is evident that the average value of function b does not depend on time.

For an equilibrium system containing a given number of particles and occupying a given volume, it is assumed that the dependence of the probability density distribution $\rho(p,q)$ on p and q takes the form

$$\rho = \rho[H(p,q)]. \tag{4.11}$$

That is, ρ depends only on the Hamiltonian function and depends in a continuous way. This statement represents the principle of equal probability of equal elements of the phase volume corresponding to a given energy.

Thus, the problem of calculating ensemble averages is reduced to the problem of obtaining an analytical representation of the probability distribution function $\rho(p,q)$ for systems with different prescribed external parameters (conditions).

Microcanonical, canonical, and grand canonical Gibbs ensembles are distinguished. In this chapter, we use only microcanonical and canonical ensembles. Therefore, these are considered later.

An *NVE* (microcanonical) ensemble is the ensemble of isolated systems. The parameters prescribed for each system are energy *E*, number of particles *N*, and volume *V*. The bulk distribution of phase points representing the states of the systems obeys the following condition:

$$E \le H(p,q) \le E + \Delta E. \tag{4.12}$$

Representative points of the systems are contained in a hyperlayer corresponding to a given energy interval ΔE, which is very small. For a system of a microcanonical ensemble

$$\rho = \text{const} = \rho_0 \text{ at } E \le H \le E + \Delta E \tag{4.13}$$

$$\rho = 0 \text{ at } H < E;\ H > E + \Delta E.$$

That is, all the states inside a given energy layer are equally probable, and those outside this layer have a zero probability. Temperature is not determined in an *NVE* ensemble.

An *NVT* (canonical) ensemble is the statistical distribution for systems that are characterized by volume *V*, given number of particles *N*, and their ability to exchange energy with the environment. The system is contained in a rigid heat-conducting case, while its energy is not bounded. It might be shown that $\rho(H)$ is determined by the equation

$$\tilde{\rho}(p,q) = A\exp\left[-\frac{H(p,q)}{\theta}\right]. \tag{4.14}$$

Hence, the probability distribution density in the phase space is an exponentially decreasing function of *H*. The quantity θ is called statistical temperature. According to Eq. 4.14, it has a dimensionality of energy

$$\theta = k_{\mathrm{B}}T, \tag{4.15}$$

where *T* is thermodynamic temperature.

The average value of function $b(p,q)$ for an *NVT* ensemble (canonical average) is calculated by the equation

$$\bar{b} = \frac{\iint b(p,q)\exp\left[-\frac{H(p,q)}{\theta}\right]dpdq}{\iint \exp\left[-\frac{H(p,q)}{\theta}\right]dpdq}. \tag{4.16}$$

4.2.3 Theory of Classical Nonideal Systems

In the simplest systems (monoatomic ideal gases and solids), interaction between molecules can be neglected. The Hamiltonian function of such systems is a sum in which each summand depends on canonical variables of a single molecule. As a result, the system energy equals a sum of energies of separate noninteracting molecules.

In real systems, it is necessary to take into account interparticle (intermolecular, interionic, etc.) interactions. In this chapter, we will consider uncharged molecules with an even number of electrons (all electrons are paired). Inert gases, liquid substances (water, methanol, acetone), and solids (aspirin, glycine). The energy of intermolecular interactions commonly equals ~10 kJ/mol. Let us assume that the interaction between ith and kth molecules depends only on the distance between them $r_{ik} \equiv r$. This is an approximation, because we neglect the polarization of electron shells and the mutual orientation of nonspherical molecules. In this approximation, the force of interaction **F** between two molecules can be presented as

$$\mathbf{F} = \mathbf{F}(r). \tag{4.17}$$

Such a force is called central force.

At large distances, neutral molecules are attracted to each other. It is known from quantum mechanics that the potential energy of interaction u is proportional to $1/r^6$. These forces are called van der Waals forces. At small intermolecular distances, molecules are repulsed from each other, so the potential energy of interaction becomes positive and goes to $+\infty$ when $r \to 0$ at $r \le b$ (the region of repulsion). It intersect the abscissa axes at $r = b$ and becomes negative at $r > b$, going to 0 at $r \to \infty$. At some point $r = d_0$, v reaches a minimum $v(d_0) = -\varepsilon$, where ε is the minimum depth.

A very rough idea of repulsion forces between molecules is given by a model of solid incompressible spheres (balls) of diameter σ. The value of σ gives the minimum distance at which molecules can approach each other. It is evident that $\sigma < d_0$. The repulsive potential quickly decreases with increasing r. It is approximately described by power functions with an exponent equal to 9 or 12.

Functions F and v in the case of central interactions are related by the relationship

$$F(r) = -\frac{dv(r)}{dr}. \tag{4.18}$$

At $r > d_0$, the force $F < 0$; that is, the potential is attractive (long-range forces). At $r < d_0$, the force $F > 0$; that is, the potential is repulsive (short-range forces).

The following components of long-range attraction forces are distinguished: orientation v_{or}, induction v_{ind}, and dispersion v_{disp} components. Dispersion interactions arise between molecules of any type, both polar (water, methanol) and nonpolar (inert gas, nitrogen, oxygen molecules). In nonpolar molecules, the dispersion contribution to the total energy of intermolecular interactions exceeds the induction and orientation contributions. The theory of dispersion interactions was developed by F. London. For more details, see Ref. [2].

Interaction between small spherical nonpolar molecules is well described by the Lennard–Jones (LJ) or 6–12 potentials

$$u = \frac{a}{r^{12}} - \frac{b}{r^6}. \tag{4.19}$$

Parameters a and b are commonly expressed through the quantity ε (the depth of the potential well in curve $v(r)$) and the parameter d_0 determining the minimum position in the curve $v(r)$

$$u = \varepsilon\left[\left(\frac{d_0}{r}\right)^{12} - 2\left(\frac{d_0}{r}\right)^{6}\right]. \tag{4.20}$$

Instead of the pair of parameters ε and d_0, parameters ε and σ are often used, where parameter σ is determined by the condition $u(\sigma) = 0$ (σ is a sum of van der Waals radii of the interacting molecules)

$$u = \varepsilon\left[\left(\frac{d_0}{r}\right)^{12} - 2\left(\frac{d_0}{r}\right)^{6}\right]. \tag{4.21}$$

Reference books commonly give values of ε and σ. For interactions between molecules A and B, the following relations are usually used:

$$\varepsilon_{AB} = \sqrt{\varepsilon_{AA}\varepsilon_{BB}}\,,\ \sigma_{AB} = \frac{\sigma_{AA} + \sigma_{BB}}{2}. \tag{4.22}$$

The term at r^{-6} can be obtained from quantum mechanics (London's dispersion law). The term at r^{-12} has no such a reliable justification; however, it gives a good and simple approximation. The potential curve has a minimum ($-\varepsilon$) at point $2^{1/6}\,\sigma$. The LJ potential reflects a superposition of two effects. The term proportional to r^{-6} describes attractive interactions. It dominates at large distances and tends to zero at $r \to \infty$. The second term, proportional to r^{-12}, describes repulsion at short distances and determines the occurrence of the steeply rising part of the curve: at $r \to 0$, $u(r)$ quickly rises to infinity. This term reflects the case that particles have finite sizes and are impenetrable to each other.

The interaction between atoms of polar molecules is commonly described by the model potential

$$u(r_{ij}) = \frac{q_i q_j}{r_{ij}} + 4\varepsilon_{ij}\left[\left(\frac{\sigma_{ij}}{r_{ij}}\right)^{12} - \left(\frac{\sigma_{ij}}{r_{ij}}\right)^{6}\right]. \tag{4.23}$$

Here, i and j designate different atoms, q_i and q_j designate partial charges. The values of parameters in Eq. 4.23 are found from quantum-chemical (QC) calculations or by using various empirical approaches.

The representation of real molecules as rigid (nonpolarizable) spherical particles reduces all the totality of interactions in the system to pair interactions depending only on r_{ij}. In this case, the potential energy U of interaction of all the molecules in the system is an additive sum of interaction energies u_{ij} of isolated pairs of molecules i and j

$$U = \frac{1}{2}\sum_{i \neq j} u_{ij} = \sum_{i<j} u_{ij}\,. \tag{4.24}$$

Equation 4.24 is approximate because the additivity of energies is fulfilled differently for interactions of different nature, namely:

- Electrostatic interaction, exactly
- Dispersion interaction, good degree of accuracy

- Induction forces, approximately
- Repulsion potential, approximately (20%)

An overwhelming number of statistical calculations performed for van der Waals systems are based on an assumption of an additive character of intermolecular interaction forces.

Note that potential (Eq. 4.23) is used to describe specific interactions, for example, hydrogen bonds whose energy can be much higher than 10 kJ/mol. This is achieved by a special selection of values of parameters that enter into the potential under consideration.

In the course of statistical thermodynamics, it is demonstrated that all the thermodynamic functions can be expressed through the statistical sum Z with the use of the canonical distribution. The direct calculation of Z for macroscopic systems is hardly realizable. It is necessary to find ways of simplifying the problem taking into account general properties of the Hamiltonian function and also invoking certain physical models, such as ideal gas or crystal models.

Consider a simplest case of monoatomic system composed of N identical particles occupying volume V. Its Hamiltonian function can be written as

$$H = \sum_{i=1}^{N} \frac{p_{xi}^2 + p_{yi}^2 + p_{zi}^2}{2m} + U(r_1, ..., r_N), \tag{4.25}$$

where p_{xi}, p_{yi}, and p_{zi} are momentum components of ith particle; r_i is its radius vector; and U is the potential energy of interaction between particles. For this system, Z is expressed as

$$Z = \left(\frac{2\pi m k_B T}{h^2} \right)^{\frac{3N}{2}} \frac{1}{N!} \int ... \int \exp\left[-\frac{U(r_1, ..., r_N)}{k_B T} \right] dr_1 ... dr_N, \tag{4.26}$$

where integration is performed over volume V.

$$Z_K = \int ... \int \exp\left[-U(r_1, ..., r_N) / k_B T \right] dr_1 ... dr_N. \tag{4.27}$$

The integral in Eq. 4.27 is taken over all configurations of the system and is named *configuration integral*.

For an ideal gas, $U = 0$, and

$$Z_K = V^N. \tag{4.28}$$

Because only the configuration integral depends on volume in Eq. 4.26, Z can be expressed as

$$Z = f(T)Z_K. \tag{4.29}$$

The quantity Z_K determines the dependence of thermodynamic functions on volume at constant T.

If the density increases, intermolecular interactions cannot be neglected, and Eq. 4.28 becomes no more correct; that is, $Z_K \neq V^N$.

Many equations were proposed to describe deviations from the equation of state of an ideal gas. The most general expression (the so-called virial expansion into a power series in the density) is given as

$$\frac{p}{k_B T} = \rho + B_2(T)\rho^2 + B_3(T)\rho^3 + \dots . \tag{4.30}$$

Here, $\rho = N/V$ is the density, while parameters B_2 and B_3 are called virial coefficients. $B_2(T)$ can be presented as

$$B_2(T) = -\frac{1}{2V}\left(Z_2 - Z_1^2\right). \tag{4.31}$$

Statistical sums Z_1 and Z_2 are presented as

$$Z_1 = \int dr_1 = V \text{ and } Z_2 = \iint \exp\left[-U_2(r_1, r_2)/k_B T\right] dr_1 dr_2 . \tag{4.32}$$

That is, to calculate $B_2(T)$, it is necessary to know the potential of pair interactions between particles in the system. In the case of an isotropic homogeneous system consisting of neutral molecules in the ground electronic state

$$U_2 = u(r),\ r = |r_2 - r_1|, \tag{4.33}$$

and $r_1 = 0$; that is, the first particle is at the origin of coordinates.

Thus, a quantitative description of a system that is not an ideal gas requires a knowledge of intermolecular interaction functions. For the calculation of the third virial coefficient, it is necessary to describe correctly three-particle interactions in the system under consideration.

A great number of analytical functions describing two-particle interactions are proposed in the literature. In current numerical calculations, these functions are called classical force fields. Analytical functions $u(r)$ that describe two-particle interactions depend on the type of the atom, its valence state, chemical environment, and other factors. Functions $u(r)$ are called atom-atom potentials, pair interaction functions, etc.

According to the method used to specify these functions, computational methods of statistical physics can be divided into two groups.

The first group is formed by approaches in which atom-atom potentials are used.

The second group includes approaches in which functions of intermolecular interaction are calculated by QC methods.

4.2.4 Partial Distribution Functions

Above we considered the main method of equilibrium statistical mechanics. Briefly, the idea of this method can be formulated as follows:

On the basis of the principle of equal prior probabilities, a certain number of equilibrium ensembles can be constructed. The canonical and grand canonical ensembles are most important of them. They become equivalent in the thermodynamic limit. Then, it is demonstrated that the normalization factors, namely, statistical sums corresponding to these ensembles, contain all information required for the calculation of thermodynamic quantities. Hence, the problem of equilibrium thermodynamics is reduced to the calculation of statistical sums.

An alternative approach to the description of an ensemble (system) of interacting particles is based on the use of distribution functions or spatial correlation functions. In this method, thermodynamic functions are expressed as the average of partial distribution functions (PDFs). This method is considerably closer to the general ideas of statistical mechanics.

Consider a certain microscopic dynamic function $b(x_1,...,x_N)$ that characterizes a many-particle system and depends on canonical variables $x_1,...,x_N$, where $x_i = (q_i,p_i)$ or $x_i \equiv q_i$. Because particles in the system are identical, it is possible to restrict themselves to the consideration of those functions in which all particles play the same part, because only such functions represent actual physical quantities. Hence, we suggest that all variables of the function are symmetrical with respect to a permutation of any variable:

$$b(x_1,\ldots,x_j,\ldots,x_i,\ldots,x_N) \equiv b(x_1,\ldots,x_i,\ldots,x_j,\ldots,x_N) \tag{4.34}$$

Any function of this type in the most general case can be presented in the following form:

$$b(x_1,\ldots,x_N) = b_0 + \sum_{j=1}^{N} b_1(x_j) + \sum_j \sum_{i=1}^{N} b_2(x_j, x_i)$$
$$+ \sum_j \sum_i \sum_{m=1}^{N} b_3(x_j, x_i, x_m) + \ldots + b_N(x_1,\ldots x_N) \tag{4.35}$$

Here, $b(x_1,\ldots,x_N)$ is a function depending only on s variables that cannot be divided into a sum of functions depending on a smaller number of variables.

To comply with Eq. 4.34, that is, to fulfill the conditions that b_s is symmetric with respect to permutation of its arguments and that b_s is the same function of any set of s variables, summation should be performed in such a way that all indices be different, which can be taken into account with the use of normalization factors, dividing each sum in Eq. 4.35 by the corresponding factor.

$$B(x_1,\ldots,x_N) = b_0 + \sum_{j=1}^{N} b_1(x_j) + (2!)^{-1} \sum_{j \neq i}\sum b_2(x_j x_i) + \ldots$$
$$+ (s!)^{-1} \sum_{j_1 \neq \ldots j_s} \sum b_s(x_{j1}, x_{j2}, \ldots, x_{js}) + \ldots + b_N(x_1,\ldots,x_N) \tag{4.36}$$

The physically real systems possess the remarkable property that dynamic functions of practical interest depend on only a finite number of irreducible functions $b_0, b_{1,\ldots,} b_s$, where $s = 2$ or 3. For such dynamic functions, $b_s \equiv 0$, if $s > r$. Hence, to average such functions, it is sufficient to know only the set of s-particle distribution functions rather than the total N-particle function, that is,

$$\langle b \rangle = \sum_{s=0}^{r} (s!)^{-1} \int dx_1 \ldots dx_s b_s(x_1,\ldots x_s) \rho_s(x_1,\ldots x_s), \tag{4.37}$$

where $r << N$, and ρ_s is the s-particle distribution function.

We will hold to the designations and the way of presentation used in works by N.N. Bogolyubov (see, for example, Refs. [3, 4]).

Inside the system, we select an arbitrary group of s particles s = 1,2,3… and define PDFs (correlative functions) in such a way that the expression

$$\frac{1}{V^s} f_s(q_1,...q_s)dq_1...dq_s \tag{4.38}$$

would determine the probability of detecting particles of this group in, respectively, small volumes dq_1, dq_2,..., dq_s in the vicinity of points q_1, q_2,..., q_s with arbitrary positions of all other particles of the system.

The multiplier V^{-s} provides (a) the dimensionless character of f_s and (b) the finiteness of f_s in the thermodynamic limit ($V \to \infty$).

It is evident that in a system of independent and noninteracting particles (ideal system)

$$f_s(q_1,...q_s) = \prod_{1 \le i \le s} f_1(q_i). \tag{4.39}$$

The differences

$$f_s(q_1,...q_s) - \prod_{1 \le i \le s} f_1(q_i) \tag{4.40}$$

characterize the existence of correlation in the positions of particles. Their absolute values can serve as a measure of ordering in the distribution of particles over positions.

The normalization condition for PDFs

$$\frac{1}{V^s} \int_V ... \int_V f_s(q_1,...q_s)dq_1...dq_s = 1 \tag{4.41}$$

and the relationship between PDFs of different orders

$$\left(\frac{1}{V^m}\right)\frac{1}{\mathrm{H}^{s\text{-}m}} \int_V ... \int_V f_s(q_1,...q_s)dq_{m+1}...dq_s = f_m(q_1,...q_m)\left(\frac{1}{V^m}\right) \tag{4.42}$$

follow from Eq. 4.39.

Equations 4.43 and 4.44 present some partial cases for Eqs. 4.41 and 4.42:

$$\frac{1}{V}\int_V f_1(q)dq = 1 \tag{4.43}$$

$$\frac{1}{\mathrm{X}_1,...\mathrm{X}_\mathrm{N}} \int_V f_s(q_1,...q_s dq_s) = f_{s-1}(q_1,...q_{s-1}), s \ge 2 \tag{4.44}$$

Consider a system that is assumed to be equilibrium and thermodynamic. Then, the function $f_N(q_1,\ldots q_N)$, where N is the total number of particles, is proportional to the configuration distribution function $D_N(q_1,\ldots q_N)$.

$$f_N(q_1,\ldots q_N) = V^N D_N(q_1,\ldots q_N) \tag{4.45}$$

$$D_N(q_1,\ldots q_N) = \int \rho_N(q,p)\frac{d^N p}{N!h^{3N}} = \frac{1}{Z_{id}}\int \exp\left[-\beta H(q,p)\right]\frac{d^N p}{N!h^{3N}}$$

$$= \frac{1}{z_{id}Q_N}\int \exp\left[-\beta H©(p)\right]\frac{d^N p}{N!h^{3N}} - \exp\left[-\beta U_N(q_1,\ldots q_N)\right]$$

$$I = \frac{1}{N!\bar{\lambda}^{3N}} = \frac{V^N}{V_N N!\bar{\lambda}^{3N}} = \frac{Z_{id}}{V^N},\ \bar{\lambda} = h/\sqrt{2\pi m/\beta};$$

$$D_N(q_1,\ldots q_N) = \frac{1}{Z_{id}Q_N}\cdot\frac{Z_{id}}{V^N}\cdot\exp\left[-\beta U_N(q_1,\ldots q_N)\right]$$

$$= \frac{1}{Q_N V^N}\exp\left[-\beta U_N\right] = \frac{1}{Q'_N}\exp\left[-\beta U_N(q_1,\ldots q_N)\right] \tag{4.46}$$

Using Eq. 4.44, it is possible to write for any f_s that

$$f_s(q_1,\ldots q_s) = \frac{V^N}{V^{N-s}} = V^s\int_V\ldots\int_V D_N(q_1,\ldots q_N)dq_{s+1}\ldots dq_N \tag{4.47}$$

Equation 4.47 at s = 1,2,3,... formally solves the problem of constructing the system of PDFs $f_s(q_1,\ldots q_s)$. In the literature, so-called generic distribution functions $\rho_s(q_1,\ldots q_s)$ are very often used. The expression $\rho_s(q_1,\ldots q_s)\ dq_1\ldots dq_s$ determine the probability that, in the system of N particles, one particle will be found in the volume element dq_1 in the vicinity of point q_1, another particle will be found in the volume element dq_2 in the vicinity of point q_2, and finally the last particle will be found in the volume dq_s in the vicinity of point q_s. The relation between f_s and ρ_s is defined by the equation

$$\rho_s(q_1,\ldots q_s) = \frac{N!}{(N-s)!}\frac{1}{V^s}f_s(q_1,\ldots q_s) \tag{4.48}$$

In fact, there are N possibilities in order that one particle be located in the volume dq_1, $(N-1)$ possibilities in order that another particle be located in the volume $dq_2,\ldots$, and $(N-s+1)$ possibilities in order that the last particle be located in the volume dq_s.

The total number of ways of realization of the state under consideration is, thus, equal to $N(N-1)...(N-s+1)=\frac{N!}{(N-s)!}$. Therefore, the normalization condition for ρ_s is written in the form

$$\int ... \int \rho_s(q_1,...,q_s)dq_1...dq_s = \frac{N!}{(N-s)!}. \quad (4.49)$$

In the thermodynamic limit,

$$\rho_s(q_1,...q_s) = \frac{N!}{(N-s)!}\frac{1}{V^s} f_s(q_1,...q_s) \approx \frac{N^s}{V^s} f_s(q_1,...q_s)$$
$$= n^s f_s(q_1,...q_s) \equiv \frac{1}{v^s} f_s(q_1,...q_s), \quad (4.50)$$

where n is the numerical concentration of particles in the system.

Along with the unconditional PDF f_s introduced above, determining the prior probabilities of positions of groups of particles, sometimes it becomes necessary to consider more complicated conditional partial distribution function (CPDF) associated with the conditional probabilities of positions of one group of particles at given positions of another group of particles.

Similarly to Eq. 4.38, let us define a CPDF in such a way that the expression

$$\frac{1}{V^s} f_s(q_1,...q_s / q_{s+1},...q_{s+p})dq_1...dq_s = \frac{1}{V^s} f_s^{\,p} dq_1...dq_s \quad (4.51)$$

signifies the probability that a selected group of s particles is found at a certain configuration under condition that the positions of another group of p particles are known in advance.

From the definition (Eq. 4.51), it follows that a CPDF can be expressed through an unconditional PDF

$$f_s(q_1,...q_s / q_{s+1},...,q_{s+p}) \equiv f_s^{\,p} = \frac{f_{s+p}(q_1,...q_{s+p})}{f_p(q_{s+1},...q_{s+p})} \quad (4.52)$$

Homogeneous systems are a very important case of systems, which correspond to real gases and liquids in the absence of forces and far from the surface (but not to solids, in particular, to crystals). A system is called spatially homogeneous if its local (intensive) physical characteristics are similar at all points of the space. Mathematically, this is expressed as translational invariance of PDFs

$f_s(q_{1+a},\ldots q_{s+a}) = f_s(q_1,q_s)$, where a is a constant vector.

Then, all the positions of one particle are equiprobable, and we obtain from Eq. 4.43 in the T limit that

$$f_1(q) = 1 \tag{4.53}$$

For homogeneous systems, the distribution of particles in vicinities of any selected particle will be spherically symmetric on the average. Therefore, the binary distribution function (BDF) $f_2(q, q')$ will depend on only one argument $\vec{r} = \vec{q} - \vec{q}'$

$f_2(\vec{q},\vec{q}') = g(|\vec{q}-\vec{q}'|) = g(|\vec{r}|) = g(r)$, where $g(r)$ is a special designation of BDFs for homogeneous systems. Then, for conditional PDFs at $p = 1$, we obtain from Eqs. 4.52 and 4.53 that

$$f_s(q_1,\ldots,q_s/q_{s+1}) \equiv f_s^1 = f_{s+1}(q_1,\ldots q_{s+1}), \tag{4.54}$$

in particular

$$f_1(\vec{q}/\vec{q}') = f_2(\vec{q},\vec{q}') = g(|\vec{q}-\vec{q}'|) = g(r). \tag{4.55}$$

Thus, the unconditional BDFs $g(r)$ have also a sense of a unary relative (conditional) distribution function.

The probability of finding a certain particle at a distance from r to $r + dr$ from another particle equals

$$d\omega(r) = g(r)\frac{4\pi r^2 dr}{V}. \tag{4.56}$$

Therefore, the function $g(r)$ is called the radial distribution function (RDF) of particles in a liquid or gas.

Using the property of homogeneity in the thermodynamic limit, Eq. 4.50 for the BDF can be written in the form

$$\rho_2(q_1, q') \equiv \rho_2(r) = n^2 f_2(q, q') \equiv n^2 g(r). \tag{4.57}$$

Because the solution for the majority of physically significant problems can be obtained using a BDF, we introduce for convenience one more function $v_2(r_{12}) \equiv h(r)$ (pair correlation function)

$$\begin{aligned} g(r_{12}) &= 1 + v_2(r_{12}) \\ g(r) &= 1 + h(r) \end{aligned}. \tag{4.58}$$

The difference between $g(r)$ and $v_2(r)$ can easily be understood if the distance between particles goes to infinity

$$\lim_{r\to\infty} v_2(r)=0 \text{ and } \lim_{r\to\infty} g(r)=1. \quad (4.59)$$

There is also a region of *excluded volume*, which is determined by parameter σ (particle diameter); see Section 4.2.3. At $r > \sigma$, a small maximum appears, and the function further quickly reaches 1.

In a liquid, the RDF is represented by a smooth curve, which exhibits maxima and minima at small values of r. The maxima correspond to most probable distances between the central particle and particles surrounding it. For a crystal, this function consists of peaks, and $g(r) \to 0$ in the intervals between peaks.

Let us consider the properties of the RDF for a liquid state in more detail. It is evident that the number of neighbors of the central molecule in spherical layers with radii r_{m1}, r_{m2}, and so on is larger than that in the case of a random distribution, so a kind of coordination sphere forms. The value of r_{m1} gives the radius of the first coordination sphere, etc. In the case of monoatomic liquid, the values of r_{m1} and r_{m2} virtually coincide with σ and 2σ. The average number of particles at a distance between r and $r + dr$ from the given particle is connected with the RDF by the equation

$$dN(r) = (N - 1)dw(r) \cong (N/V)g(r)\pi r^2 dr. \quad (4.60)$$

The average number of molecules contained in the first coordination sphere, that is, at distances between r_{m1} and r_{m2} from the given molecule, is called the coordination number. For example, in water at temperature $T = 1.5$ C (slightly above the freezing point), $r_{m1} = 2.9$ Å and $r_{m2} = 4.5$ Å, while the coordination number is $z = 4.4$. For comparison, in water ice crystals (ice 1 h), $r_{m1} = 2.76$ Å, $r_{m2} = 4.51$, and $z = 4.4$.

The coordination number can be calculated by the equation following from Eq. 4.60

$$z = \int_{r_{M1}}^{r_{M2}} (N-1)dw(r) = 4\pi\frac{N}{V}\int_{r_{M1}}^{r_{M2}} g(r)r^2 dr. \quad (4.61)$$

For many liquids near the melting point, r_{m1} is very close to the distance between the nearest neighbors in the crystal. Thus, from the values for water and ice given before, it is seen that the radii of coordination spheres and the number of molecules in the first coordination sphere in water at low temperature are only slightly differ from those for the crystal.

In liquids, there is a finite probability that a particle can be found at any distance $r > r_{M1}$ from the given molecule. This fact leads to fluctuations in coordination numbers. Therefore, it is necessary to stress that the order in liquids and crystals is of fundamentally different character.

4.2.5 Integral Equations of the Theory of Liquids: Direct Correlation Function

The calculation of z (see Eq. 4.26) for liquid systems is *practically impossible*. Therefore, starting in 19th century, alternative approaches to the quantitative description of properties of liquids had been being developed. These approaches are based on paired correlation functions, while RDF are found analytically by solving integral equations of the theory of liquids.

Consider a system of N identical particles contained in volume V. Let us suppose that interaction forces acting between particles are central ones. Previously, we introduced the RDF $g(r)$ (see Section 4.2.5). The quantity $[g(r) - 1]$, decreasing to zero with increasing r, is called the pair correlation function. The most fundamental property of this function is expressed by the equation

$$1+\rho\int[g(1,2)-1]d2 = k_{\mathrm{B}}T\left(\partial\rho/\partial p\right)_T, \tag{4.62}$$

where $g(r) \equiv g(1,2)$. This equation connects the spatial integral of $[g(1,2) - 1]$ with the compressibility of the liquid.

Thermodynamic and other macroscopic parameters of the system are unambiguously determined by the set of correlation functions. Thus, the knowledge of correlation functions allows one to determine the fundamental relation between the structure and the properties of the liquid. To construct a structural theory of the liquid, it is necessary to know a method for the calculation of $g(r)$ from $u(r)$. The system of Bogolyubov–Green–Kirkwood–Born equations provides, in principle, such a possibility. This is an infinite system of engaging integro-differential equations for N-particle distribution functions. In this system, the one-particle correlation function is connected with the two-particle one, and so on. As the starting point, it is possible to take the equation for $g(1,2)$ in the form of the ratio of two configuration integrals of Boltzmann factors

$$g(1,2) = \frac{V^2 \int \exp\{-U(1,2,...,N)/k_B T\} d3...dN}{Z_k}. \tag{4.63}$$

All integrals are taken over the volume V, and Z_k is the configuration integral (see Eq. 4.27). Here, we will use abbreviations $1 \equiv r_1$, $d3 \equiv dr_3$, etc. The expression

$$w(1,2) = \frac{\exp\{-U(1,2,...,N)/k_B T\} d3...dN}{Z_k} \tag{4.64}$$

is the probability that the first particle is contained in a volume element $d1 = dr_1$ in the vicinity of the point with the radius vector r_1, and the second particle is contained in the volume element dr_2 in the vicinity of the point with the radius vector r_2 irrespective of the position of all other $N - 2$ particles, whose coordinates can take arbitrary values. Then, it is assumed that the interaction between particles is pairwise and is represented in the form

$$U(1,2,...N) = u(1,2) + \sum_{j \geq 3} u(1,j) + . \tag{4.65}$$

Differentiating Eq. 4.65, one can obtain

$$-k_B T \nabla_1 g(1,2) = g(1,2) \nabla_1 u(1,2) + \rho \int g(1,2,3) \nabla_1 u(1,3) d3. \tag{4.66}$$

Here, ∇_1 means the gradient with respect to coordinates of molecule 1, and $g(1,2,3)$ is a three-particle correlation function.

In the framework of the approximation of pair interactions, there exists a coupled closed system of two integral equations for two-particle (Eq. 4.66) and one-particle (Eq. 4.67) distribution functions, which is called the Bogolyubov–Green–Kirkwood–Born equation system

$$-k_B T \nabla_1 w(1) = \int w(1,2) \nabla_1 u(1,2) d2. \tag{4.67}$$

Here, $w(1)$ is the probability that the first particle is contained in the volume element $d1 = dr_1$ in the vicinity of point with the radius vector r_1 irrespective of the positions of other $(N - 1)$ particles, $w(1,2)$ is defined in Eq. 4.66.

Equation 4.66 is an exact one and connects $g(r)$ and $u(r)$ with each other. However, it contains one more unknown function $g(1,2,3)$. Kirkwood proposed to present the function $g(1,2,3)$ in the form

$$g(1,2,3) = g(1,2)\, g(1,3)\, g(2,3). \tag{4.68}$$

This is the so-called superposition approximation. It has no physical justification and represents a simplification of a mathematical character.

The system of the two Eqs. 4.66 and 4.67 is one of the main instruments in studying the structure and properties of liquids. In the absence of an external field, this system turns into the Ornstein–Zernike equation for the pair correlation function $h(r)$

$$h(r) = g(r) - 1. \tag{4.69a}$$

Here, $g(r)$ is the RDF function introduced in Section 4.4. It is evident that the function $[g(r) - 1]$ tends to zero at $r \to \infty$. Passing from the system of coordinates in which particle 1 is at the origin of coordinates to an arbitrary system of coordinates, we will write down Eq. 4.69a in a different way as

$$h(r_{12}) = g(r_{12}) - 1. \tag{4.69b}$$

Hence, the function $h(r_{12})$ is a measure of the total effect of molecule 1 on molecule 2 at a distance of r_{12}. Ornstein and Zernike presented $h(r_{12})$ as

$$h(r_{12}) = c(r_{12}) + \rho \int c(r_{13}) h(r_{23}) d\mathbf{r}_3 . \tag{4.70}$$

The first term on the right-hand side of Eq. 4.70 is the direct correlation function, and the second term describes the effect of particle 1 on particle 3. The latter, in its turn, acts on particle 2 either directly or indirectly (through other particles). The direct correlation function $c(r_{12})$, as distinct from the RDF, has no clear physical meaning. However, it is much simpler. It disappears much faster than $h(r)$. The tail of the last-mentioned function is almost completely determined by the second term on the right-hand side of Eq. 4.70.

The direct correlation function and the Ornstein–Zernike Eq. 4.69 form the basis of methods that are suitable for the description of a liquid state. As distinct from the system of Bogolyubov–

Green–Kirkwood–Born equations, in which the superposition approximation (Eq. 4.67) is used, the Ornstein–Zernike equation is exact.

Integral equations are widely used for the verification of phenomenological models of liquid. Their applicability is restricted by (i) problems in finding pair interaction functions and (ii) their inapplicability to the description of metastable and amorphous states, because the equations are obtained in the thermodynamic limit.

4.2.6 Use of the PDF Apparatus for Calculations of Thermodynamic Characteristics of Nonideal Systems

All the thermodynamic quantities, which are initially introduced as the averages of dynamic functions over the phase space, can be very easily expressed through PDFs. The problem is simplified due to the independence of momenta and coordinates in the Hamiltonian function $H(q,p)=\sum_{i}^{N}\frac{p_i^{\,2}}{2m}+H_N(q_1,...q_N)$.

Therefore, many thermodynamic quantities can be presented in the form $X = X^{\text{kin}} + X^{\text{pot}}$. Here, we give expressions for a number of macroscopic characteristics of nonideal systems.

Internal energy $E(T,N,V)$:

$$U \equiv E = \int H(q,p)\rho_N(q,p)d^N q d^N p$$

$$E = \int H^{\text{kin}}(p)\rho_N(p)\frac{dp_1...dp_N}{N!h^{3N}}$$

$$+\int U_N(q_1,...q_N)\rho_N(q_1,...q_N)dq_1...dq_N$$

$$E = E^{\text{kin}} + E^{\text{pot}}$$

$$E^{\text{kin}} = \frac{3}{2}NkT$$

$$E^{\text{pot}} = \int U_N(q_1,...q_N)\rho_N(q_1,...q_N)dq_1...dq_N = 2\pi n^2 V\int_0^{\infty} V_{12}(r)g(r)r^2 dr$$

$$E = \frac{3}{2}NkT + 2\pi n^2 V \int_0^\infty V_{12}(r) g_{12}(r) r^2 dr = N\frac{3}{2}kT + \frac{2\pi}{v} \quad (4.71)$$

$$\int_0^\infty V_{12}(r) g(r) r^2 dr$$

Pressure:

$$P = \frac{N}{V}kT - \frac{2\pi n^2}{3}\int_0^\infty \left(\frac{dV(r)}{dr}\right) drr^3 dr \quad (4.72)$$

Isothermal compressibility:

$$k_B T\left(\frac{\partial n}{\partial P}\right)_T \equiv nkT\chi_T = 1 + n\int d\vec{r}v_2(r)$$

$$nkT\chi_T = 1 + 4\pi n\int_0 r^2 (g(r) - 1) dr$$

$$\chi_T = \frac{1}{nkT}\left\{1 + 4\pi n\int_0^\infty r^2 v^2(r) dr\right\} \chi_T = -\frac{1}{V}\left(\frac{\partial V}{\partial P}\right)_T \quad (4.73)$$

4.3 Basics of Classical MD Simulation

4.3.1 Computer Simulation of Macroscopic Properties

Quantum chemistry allows one to calculate *microscopic* properties of molecules in the gas phase or in an ideal crystal (see Chapter 3). These properties include structural parameters, spectra, bond energies, etc. For the calculation of macroscopic properties (virial coefficients, diffusion characteristics, heat capacity, etc.), it is necessary to consider molecular ensembles (see Section 4.2). In the computer implementation of the method of Gibbs ensembles, it is necessary to answer some questions, in particular:

- What is the minimal number of particles required for the simulation of macroscopic properties?
- What is understood under the term «particle»? Is that an atom, a fragment of a molecule, or a molecule as a whole?

The first question can be reformulated as follows: Does crystallization exist in the system of rigid balls the interactions between which are reduced to only an excluded volume? In 1957, B. J. Alder and T. E. Wainwright first used the method of MD for studying the properties of a system that consisted of 32 rigid spheres in a cubic cell with periodic boundary conditions [5]. They observed a first-order phase transition between liquid and solid phases (crystallization). Simultaneously, W. W. Wood and J. D. Jacobson [6] obtained the same result by the Monte Carlo (MC) method. This example shows that (i) certain macroscopic properties can be simulated with a small number of particles (the linear dimensions of the cell were about 1 nm) and (ii) particles can have only remote relation to atoms or molecules. It is evident that the number of particles in the cell and the model used for the description of particles are determined by the goals of computer simulation.

The notion of interacting particles is fundamental in computer simulation. It is called *atomic sites*. Each particle of the system is a discrete object characterized by the position in the space (its coordinates **r** and momentum **p** in the MD method or only vector **r** in the MC method) and also by the interaction potential **u(r)** with other objects (see Section 4.2.3). Various objects can serve as particles: (i) separate atoms, (ii) functional groups (carbonyl C=O, hydroxyl O–H, methyl CH_3, etc.), (iii) small solvent molecules (water), and (iv) atomic groups (repeating links in a polymer chain, amino acid residues).

The key role in computer simulation is played by the interaction potential **u(r)**. In the case of conservative systems (the only case considered here), the notions of *force* and *potential* are equivalent. The methods of computer simulations can be divided into the following two groups according to the way of obtaining interaction forces/potentials:

- MD methods with ab initio force fields. In these methods (Born–Oppenheimer molecular dynamics [BOMD] and Car–Parrinello molecular dynamics [CPMD]), the motion of nuclei is described by the classical equations of motion, while forces acting on nuclei are found by solving the electronic Schrödinger equation.

- MD and MC methods with classical force fields. In these methods pair potentials expressed through analytical functions are used. These methods are based on a mechanical model of a molecule; that is, a molecule is treated as a system of bonded classical particles with definite masses and charges (see Eq. 4.3) rather than a collection of electrons and nuclei obeying the laws of quantum mechanics.

  A certain type of particles used in calculations and the method of computer simulation are selected on the basis of the spatial and timescales required for the calculation of given macroscopic properties of the system under consideration.
- MD methods with ab initio force fields are used in the calculations of systems that contain several hundreds of particles (a cell of ~1 nm) at times of tens picoseconds, for example, for the estimation of the proton diffusion coefficient in acidified aqueous solutions (see Ref. [7]).
- MD and MC calculations with classical pair potentials are performed for systems containing several thousands of particles (a cell of ~10 nm) at times of hundreds of nanoseconds (~10^{-7}), for example, in studying conformational changes in DNA fragments (see Ref. [8]).
- In combined approaches, the system under consideration is divided into a quantum and a classical regions, so the chemical subsystem is calculated maximally accurate, while approximate methods are used in the description of environment. These methods are widely used in the simulation of processes/reactions in biosystems and in solutions (see Refs. [9, 10]).
- Coarse-grained models are used for cells of ~10 nm and times of tens of microseconds (~10^{-5} s), for example, in studying the interaction of polymers with the substrate. In this case, interactions between amino acid residues are considered (see Ref. [11]).
- The method of integral equations reference interaction site model (RISM) theory of Curro and Schweizer [12, 13]. This method is based on the solution of integral equations and the formalism of correlation functions (Section 4.2.5) and is used in the simulation of properties of liquids and reactions in the liquid phase (see Ref. [14]).

The methods described here use interaction potentials of particles, that is, they give a microscopic description of the simulated system. There is a completely different class of models that treat the simulated system in terms of collective variables, namely, potential fields and densities. These approaches will not be considered in this book.

We use here the term "computer simulation." This term is equivalent to the term "computer experiment," introduced by L. Verlet in 1967 [15]. Note that this term in the original paper by Verlet is written in quotes. Subsequently, the quotes were skipped, which resulted in certain indefiniteness in the presentation of the MD and MC methods. It was not clear which kind of experiment, physical or chemical, is the case in point. In a physical experiment, the response of the studied object either to variations of parameters (temperature, density, composition) or to external effects is commonly investigated. In this case, the rejection of the consideration of a molecule as a collection of nuclei and electrons is in the majority of cases quite justified.

In a chemical experiment, conversion of some molecules into some other ones is commonly investigated. This conversion is accompanied by the breaking of some covalent bonds and the formation of some other bonds (organic, inorganic, metalorganic synthesis). The rejection of the consideration of a molecule as a collection of nuclei and electrons is absolutely not justified in this case. In other cases, for example, in studying the formation or breaking of noncovalent bonds (simulation of the structure of proteins, supramolecular synthesis), the use of a mechanical model of a molecule is quite justified. In our opinion, it is desirable to use the term "computer experiment" only for physical experiments.

4.3.2 Comparative Characteristics of Simulation Methods for Disordered Condensed Systems

Currently, MD and MC methods with classical pair potentials and MD methods with ab initio force fields (BOMD and CPMD) are most widely used in the computer simulation of disordered condensed systems.

Problems with the sizes of the system that is explicitly taken into account in calculations, that is, the shape of the cell and the number of particles in the cell, are a common deficiency of these methods. It is evident that, in the case of using classical pair potentials, the sizes of the cell are several orders of magnitude larger than the sizes of the cell in CPMD calculations. However, these sizes are still not sufficiently large in order that the structure and collective conformational transformations in protein globules could be correctly simulated.

The length of the step, which is ~1 fs in MD and MC methods and 0.1 fs in CPMD, essentially restricts the length of the trajectory. The length of the trajectory in MD calculations is commonly not larger than 1 μs (10^9 steps) and 20 ps (10^5 steps) in CPMD. It is evident that processes with characteristic times of 1–10 ms, for example, collective conformational reorganizations that involve regions of the secondary structure and subunits of protein globules.

The possibility of calculations of both microscopic (correlation density-density function, vibrational spectra, chemical nuclear magnetic resonance shifts) and macroscopic (diffusion and virial coefficients) systems in a condensed state (liquids, crystals, structured media, etc.) at different values of temperature and pressure. Calculated characteristics can be compared with available experimental data or values obtained using approximate equations of the theory of liquids. In the case of methods based on classical pair potentials, the accuracy of the obtained values is determined by the type of the force field used in the calculations. Note that a great number of force fields that reproduce well a certain property of a certain system are proposed by now. If the simulated system is not large, and the calculated property does not require obtaining relatively long trajectories, interactions in the system can be calculated rather accurately by density functional theory (DFT) methods. The characteristics obtained in this way commonly exhibit a quantitative agreement with available experimental data.

4.4 The Monte Carlo Method

The application of rigorous methods of statistical physics to classical condensed systems meets very serious difficulties because of the necessity of calculation and asymptotic estimation (at $N,V \to \infty$) of the configuration integral (Eq. 4.27). The main idea of the MC method consists of the replacement of direct multiple integration in Eq. 4.26 or in similar integral expressions that determine the average value of function M of coordinates by numerical integration over a random set of points (with certain special rules of selection of these points)

$$\langle M\rangle=\frac{1}{Z_K}\int\ldots\int M(r_1,\ldots,r_N)\exp\{-U(r_1,\ldots,r_N)/kT\}dr_1,\ldots dr_N. \quad (4.74)$$

In numerical calculations, integration is commonly replaced by summation. Let us divide the $3N$-dimensional configuration space of the studied system into an arbitrarily large number s of numbered cells of equal volume. We will say that the system occurs in the ith state if its representation point occurs in the ith cell. In each state, we can assign a certain numerical value M_i to any function of the coordinates of the system $M(r_1,\ldots,r_N)$ taking values of $r_1,\ldots,r_N$ corresponding, for example, to centers of cells. The interaction energy of system particles will be now represented by a set of its possible values U_i, $i = 1,2,\ldots,s$. If s is sufficiently large, the replacement of the continuous configuration space by a discrete one will not practically affect the estimation of the average values of functions of coordinates. Then, instead of Eqs. 4.26 and 4.74, we will have

$$Z_K\approx Z_K^*=\sum_{i=1}^{s}\exp\{-U_i/kT\} \quad (4.75)$$

and

$$\langle M\rangle=\frac{1}{Z_K^*}\sum_{i=1}^{s}M_i\exp\{-U_i/kT\}. \quad (4.76)$$

Let us present now the following formal scheme, which is called homogeneous Markovian chain. Consider the set of all s possible states of the system as a collection of random events A_i that form a sequence (Markovian chain) with constant probabilities of transitions $A_i \to A_j$ equal to $p_{ij} \geq 0$ and obeying the normalization condition

$$\sum_{j=1}^{s}p_{ij}=1; j=1,2,\ldots,s. \quad (4.77)$$

Let us designate the probability that the transition $A_i \to A_j$ occurs in n steps as p_{ij}^n (so that, $p_{ij}^n \equiv p_{ij}$). If all A_i, $i = 1,2,\ldots, s$, form one ergodic class (i.e., all states A_i are not periodic, and any state A_j is attainable from any state A_i for a certain finite number of transitions n), the limiting transition probabilities exist

$$\lim_{n\to\infty} p_{ij}^{(n)} = \omega_j, j = 1,2,\ldots s, \tag{4.78}$$

where

$$\omega_j > 0, \sum_{j=1}^{s} \omega_j = 1 \tag{4.79}$$

for all i, so ω_j realize a certain distribution of probabilities for A_j. In addition, in the theory of Markovian chains, it is proved that, if the normalization condition (Eq. 4.79) is obeyed, the values of ω_j are unambiguously determined by the values of p_{ij} from the system of linear equations

$$\omega_j = \sum_{j=1}^{s} \omega_i p_{ij}, j = 1,2,\ldots,s \tag{4.80}$$

and that the distribution determined by numbers ω_j is a stationary distribution of probabilities of events A_j; that is, if this distribution would be taken as an initial one, it would not change during the studied Markovian process.

Equation 4.80 expressed the tendency of the system to approach a stationary state regardless of the initial state selected. Therefore, if a chain with an unrestricted number of steps is considered, the average value of a certain function of state taken along the chain will be evidently equal to

$$\langle M \rangle = \sum_{i=1}^{s} M_i \omega_i, \tag{4.81}$$

where

$$\omega_i = \frac{1}{Z_K^*} \exp\{-U_i / kT\}. \tag{4.82}$$

The limiting average of M_i along an unrestricted Markovian chain coincides with the canonical average M according to Eq. 4.76. Now, it is remained to select the transition probabilities p_{ij}. This can be done by many ways, because s^2 quantities p_{ij} are not determined uniquely from $2s$ Eqs. 4.80 and 4.77. For this purpose, relationships of the following form are introduced:

$$p_{ij} \exp\left(-\frac{U_i}{kT}\right) = p_{ji} \exp\left(-\frac{U_j}{kT}\right) \quad (4.83)$$

Thus, conditional transition probabilities are introduced: the probability of a given event, which consists of the appearance of configuration j, depends on the preceding event.

Hence, a Markovian chain all states of which form one ergodic class and the transition probabilities of which obey Eqs. 4.77 and 4.83 converges to a canonical Gibbs ensemble in the sense that different states tend to appear with frequencies proportional to Boltzmann factors $\exp\{-U_i/kT\}$ at sufficiently long continuation of the chain length. The convergence does not depend on the choice of the initial state, and the averages of the functions of states along the chain tend to average over the canonical ensemble. In 1953, Metropolis et al. [16] proposed a method for the numerical computation of the canonical averages such that the frequency ω_i of the appearance of an arbitrary configuration was proportional to $\exp[-U_i / kT]$

$$\omega_i = \lim_{L\to\infty} \frac{L_i}{L} = \exp\left(-\frac{U_i}{k_B T}\right) \Big/ \left[\sum_{i=1}^{S} \exp\left(-\frac{U_i}{k_B T}\right)\right]. \quad (4.84)$$

Here, L is the total number of trials, L_i is the number of trials in which the ith configuration was observed with the probability ω_i. In the realization of such a chain for a system of hard nonattracting balls, all the configurations in which at least two balls overlap are *excluded*, while all other configurations are *equiprobable*.

The idea of reducing canonical averaging to averaging along a certain Markovian chain using the Metropolis algorithm is implemented in computer programs by reproducing the corresponding Markovian chains. Such algorithms are convenient for programming. The practical implementation of Markovian chains is based on the construction of one-step transition probabilities, a detailed description of which is outside the scope of this book.

The calculation by the MC method consists of obtaining numerical values of macroscopic characteristics at given conditions. For the canonical *NVT* ensemble (Section 4.2), the calculation gives a point on the diagram of the dependence of the parameter $\langle M\rangle$ on the variables T and V/N. Note that all the calculations for systems with a different interaction potentials must be started anew. The calculations are performed for a cubic cell of size V containing N particles. The interaction energy between particles i and k is described by a certain potential $u(r_{ik})$. Periodic boundary conditions are imposed on the system, and a cutoff radius r_{cut} is used in calculations (see Section 4.7). One of the possible calculated characteristics is the RDF (see Section 4.2.5). To calculate the RDF, the number of molecules $N(r)$ around a given molecule contained inside a sphere of radius R is required to be determined. The values of $N(r)$ are averaged over different molecules and along the chain of configurations. The accuracy of calculations by the MC method can be increased by increasing the number of statistical trials.

4.5 Methodology of Molecular Dynamics Simulations

The classical method of MD starts from the presentation of a real molecular object by a mathematical conception, namely by a system of interacting material points (particles, atomic sites), whose motion is described by classical Newton equations (for more details, see Chapter 2).

$$m_i \frac{d^2 \mathrm{r}_i}{dt^2} = \mathrm{F}_i, i = 1,\ldots N \tag{4.85}$$

Here, N is the number of particles in the system, m_i are their masses, and $\mathbf{F}_i$ is the force acting on a particle with number i. Consider a very small interval of time δt. During this time, the force $\mathbf{F}$ can be considered constant. The position of any particle $r_i(t)$ and its velocity $\mathbf{v}_i(t)$ change insignificantly during time δt. With regard to Newton's law (Eq. 4.104)

$$\mathbf{v}_i(t+\delta t) \approx \mathbf{v}_i(t) + \delta t \frac{\mathrm{F}_i}{m_i} \text{ and } r_i(t+\delta t) \approx r_i(t) + \mathbf{v}_i(t)\delta. \tag{4.86}$$

The accuracy to which these approximate equations are obeyed will be the higher the smaller is the time step δt. Let us assume that we know how to calculate forces acting on the particles and that these forces depend only on the mutual positions of the particles, that is,

$$\mathrm{F}_i = \mathrm{F}_i(\mathrm{r}_1,\ldots,\mathrm{r}_N),\ i = 1,\ldots, N. \tag{4.87}$$

Then, if the coordinates and velocities of all the particles at an initial instant of time are specified, then the coordinates and velocities of each particle at the instant of time δt can be found by Eq. 4.85. Hence, it is possible to follow the behavior of a molecular system corresponding to the selected initial data and to the given forces of interaction between the particles.

Classical equations of motion (Eq. 4.85) have the first integrals: in the process of the evolution of the system, its total energy, momentum, and angular momentum are conserved. Eventually, the molecular system comes to a certain equilibrium state in which its kinetical and potential energies fluctuate near their average values, which are functions of the total energy. If the number of particles is sufficiently large, it is natural to characterize the state of the system by its temperature. The notion of temperature itself suggests interaction with an outer medium or thermostat. In this case, temperature is uniquely connected with the average (over the ensemble of states) kinetic energy of the particle. For any particle, it average kinetic energy is independent of either the mass of the particle nor the potentials of its interaction with surrounding particles and equals $(3/2)k_BT$, where k_B is the Boltzmann constant. Let us define the instant temperature $T(t)$ as

$$k_BT(t) = 2K(t)/s. \qquad (4.88)$$

Here $K(t)$ is an instant (at the instant of time t) kinetic energy, and $s = 3N - 6$ is the total number of degrees of freedom of the system. Prescribing the coordinates and velocities of all particles at the initial instant of time, we thereby also prescribe the total energy of the system $E = U(0) + K(0)$, which, by virtue of the equations of motion, will be conserved in the course of the evolution of the molecular system. The given energy will correspond to a certain temperature, which can be found from the trajectory of the system (this question will be considered later).

Computer MD simulations can be divided into three parts:

- Initialization
- Attainment of an equilibrium (thermostating or equilibration)
- Calculation of the trajectory

The first stage of MD simulation (initialization) consists of the construction of a cell and the assignment of initial conditions. In the simplest case of a homogeneous liquid, the volume of the cell containing N particles is found from the experimental density of the liquid. The initial positions of particles are obtained by finding a potential energy minimum for the system of N particles. The initial momenta are found by the equation

$$p(\mathbf{v}_{ix}) = \left(\frac{m_i}{2\pi k_B T}\right)^{1/2} \exp\left[-\frac{1}{2}\frac{m_i \mathbf{v}_{ix}^2}{k_B T}\right], \tag{4.89}$$

the velocities $\mathbf{v}_{ix}$ are obtained from the Boltzmann distribution at temperature T, which gives the probabilities that atom i with mass m_i has velocity $\mathbf{v}_x$ in the direction x. At the initial instant of time, the velocities are corrected in such a way that the total momentum is equal to zero:

$$P = \sum_{i=1}^{N} m_i v_i = 0 \tag{4.90}$$

Note that the choice of correct initial conditions is necessary to avoid the divergence of the trajectory.

The second stage of MD simulation (thermostating) is necessary to transfer the system to the equilibrium state. This is achieved by adding or taking off energy. This task is fulfilled with the use of thermostat (see later) by decreasing or increasing the kinetic energy. Upon integrating the equations of motion, the system relaxes to an equilibrium state. Two problems arise at this stage:

- The relaxation time of the system can be too long. The time step determines the real time of computer simulation. If the internal relaxation time is long, many steps are required for the system to attain an equilibrium. Various approaches are developed currently to avoid this problem. For a system studied in the region of phase transitions, a probability exists that it falls into a metastable state.
- The system can be initially positioned outside the considered part of the phase space. To solve this problem, it is necessary to perform simulations with various initial conditions and during various periods of time.

At the third stage of simulation (calculation of the trajectory), characteristics of the system are calculated. These characteristics will be considered in Sections 4.5.7 and 4.5.8.

In an experiment, molecules of interest are commonly contained in solutions and actively interact with each other or with solvent molecules. The temperature of the system is maintained by means of energy exchange with the environment. Taking into account the interaction between a molecule and the environment in detail is often impossible. To take into account the effects of energy exchange with the environment, special algorithms called are used, which are named *thermostats*. The use of a thermostat is especially important at the second stage of MD simulation. Later, we will consider two most widespread models of thermostats: (i) the collision thermostat based on collision dynamics and (ii) the Berendsen thermostat using alternating nonlinear friction in equations of motion.

In the collision thermostat model, a medium of virtual particles interacting with particles of the studied molecular system is introduced. Collisions obey the law of elastic balls. The best agreement with experimental data is achieved by varying the mass of virtual particles and the frequency of collisions with atoms of the system. For calculations in vacuum, it is common to set the mass of virtual particles at 18 aem and the frequency of collisions at 55–60 ps^{-1}. This medium is close to water in its viscosity at normal conditions. The thermostat temperature determines the velocity distribution function of virtual particles as follows:

$$f(v)=\left(\frac{m_v}{2\pi k_B T}\right)^{3/2} v^2 \exp\left(-\frac{m_v v^2}{2k_B T}\right), \tag{4.91}$$

where $f(v)$ is the probability distribution function of virtual particles over velocities ($f(v)dv$ is the probability that the absolute value of the velocity of a virtual particles is in the interval from v to $v + dv$), and m_v is the mass of a virtual particle. The frequency of collisions is commonly presented by the Poisson distribution.

The Berendsen algorithm is based on the introduction of an alternating friction. The deviations of the temperature from its equilibrium value (T_0) are corrected according to the Landau–Teller equation

$$\frac{dT(t)}{dt}=\frac{T_0 - T(t)}{\tau}. \tag{4.82}$$

Temperature deviations decrease exponentially with characteristic time τ.

4.6 Numerical Methods of the Solution of Equations of Motion: The Verlet Algorithm

In the MD method with classical force fields, the motion of a particle with mass m is commonly described by the Newtonian equation

$$F = ma = m\frac{dv}{dt} = m\frac{d^2r}{dt^2}. \tag{4.93}$$

If the acceleration a = const, we obtain an expression for r at t as a function of acceleration a, initial velocity $\mathbf{v}_0$, and initial position x_0

$$r = (1/2)at^2 + v_0t + r_0. \tag{4.94}$$

With a known force (or potential energy), we can calculate a at point r as

$$a = -\frac{1}{m}\frac{dU_N}{dr}. \tag{4.95}$$

Equations 4.93–4.95 allow the trajectory to be calculated if the initial positions of particles and their velocities are known. In the integration of equations of motion, the positions of particles, their velocities, and their accelerations are presented in the form of a Taylor series:

$$r(t+\delta t) = r(t) + v(t)\delta t + (1/2)a(t)\delta t^2 + ... \tag{4.96}$$

$$v(t+\delta t) = v(t) + a(t)\delta t + (1/2)b(t)\delta t^2 + ...$$

$$a(t+\delta t) = a(t) + b(t)\delta t + ...$$

A great number of methods have been proposed for the numerical solution of Eqs. 4.96, such as the Verlet algorithm, the Leap–Frog algorithm, etc. Later, we will consider the Verlet algorithm. To derive this algorithm, let us write down two equations:

$$r(t + \delta t) = r(t) + v(t)\delta t + (1/2)a(t)\delta t^2 + \ldots$$

$$r(t - \delta t) = r(t) - v(t)\delta t + (1/2)a(t)\delta t^2 + \ldots$$

Summing up the two latter equations, we obtain

$$r(t + \delta t) = 2r(t) - r(t - \delta t) + a(t)\delta t^2. \tag{4.97}$$

For the calculation of the positions of particles at the instant of time $t + \delta t$, the Verlet algorithm (Eq. 4.97) uses the positions and accelerations of particles at the instant of time t and positions of particles at the instant of time $t - \delta t$. Note that velocities do not explicitly enter into Eq. 4.97.

The Verlet algorithm has the following advantages: its programming is rather simple, the energy is virtually constant, and the local error is of order $O(\delta t^4)$. The disadvantage of this algorithm is the inconvenient treatment of velocities.

Because of the effect of numerical errors, the trajectories of particles obtained by the MD method using the Verlet algorithm in fact obey the equation

$$d^2x / dt^2 = F(x) + z(t). \tag{4.98}$$

Here, F is the force acting on a particle with mass equal to 1; equations are written for the projection onto axis x, all indices are skipped.

The presence of additional items in equations of motion makes the system nonconservative. In MD methods, the item z must be organized so that either its action would not give rise to energy drift or this drift would be negligibly small. In this case, the total energy can fluctuate around a certain average value. In order that energy oscillations with respect to the average value would be of stationary character, it is necessary that the work of force z in average should be equal to zero:

$$\langle z \times dx/dt \rangle = 0 \tag{4.99}$$

Because $z(t)$ is expressed through derivatives of x, Eq. 4.99 is obeyed only in the case when $z(t)$ contains only even derivatives of x. As applied to difference schemes, the aforesaid means that the error in the approximation of the second derivative must be even with respect to time. It was mentioned above that the local error in the Verlet algorithm is of order $O(\delta t^4)$, that is, contains only items that are even with respect to time. That is, the energy is virtually constant.

The total energy of the system in the course of MD calculations remains constant $\langle E \rangle$ on the average but undergoes fluctuations because of computational errors. That is, the energy is not an integral of motion. MD calculations showed that the average energy fluctuation could be estimated as

$$\Delta\left\langle E2 \right\rangle = vNk_{\mathrm{B}}Tmv_0^2 \ . \tag{4.100}$$

Thus, the realized MD ensemble differs from traditional ensembles of statistical physics. The accuracy of the calculated averages is provided by the conservation of the total energy in the average.

4.7 The Notion of an MD Cell: Periodic Boundary Conditions

In MD calculations, the 3D space is divided into equal cells of volume V with N particles in each. Commonly, cubic cells with the edge L are used. One of them is conditionally taken as the main one (MD cell). The relative configuration of particles and the changes of their coordinates in transitions are repeated in all other cells, so the displacement of the entire system by the distance of the edge length L of the main cell along this edge makes no changes in the total configuration of the system.

If one of the particles of the MD cell comes out of this cell through a certain side, its image in a neighboring cell enters into the main cell through the opposite side and occupies a position in the MD cell. Similar displacements occur in all cells. The MD cell contains N particles as before. That is, even though the systems is unrestricted, it is necessary to follow only coordinates of N particles in the MD cell, because these coordinates specify the coordinates of particles of the entire system. When periodic boundary conditions are introduced, it is assumed that the calculation of the energy of any configuration of particles in the main cell is performed by summation with all particles of the entire system, so there are no boundary effects at cell boundaries.

Consider a system consisting of spherical nonpolar molecules whose interaction energy is represented as a sum of pair interactions (see Eqs. 4.21 and 4.22). In this case, the interaction energy per an MD cell can be written as

$$U_N(r_{ij}) = \frac{1}{2}\sum_{i=1}^{N}\sum_{j=1,j\neq i}^{\infty} u(r_{ij}), \tag{4.101}$$

where $u(r_{ij})$ is the pair interaction potential (see Section 4.2.3). Index i numbers particles of the MD cell, while index j numbers particles of the whole system, the MD cell included. Approaches that allow a finite value of the limit in the internal cycle of expression to be introduced are based on so-called cutoff radii. The potential energy of the interaction of a given particle with its environment decreases with r according the law

$$u(r) \sim r^{-\alpha},\ \alpha \geq 1. \tag{4.102}$$

It is evident that, at $r > r_{\text{cut}}$, where $r_{\text{cut}} \gg \sigma$ (σ is the van der Waals radius of the particle; see Section 4.2.3). In calculations of the interaction energy of an arbitrary particle in a cell, it is necessary to consider particles in the MD and neighboring cells. The value of the cutoff radius commonly varies within some range

$$3\sigma < r_{\text{cut}} < 10\sigma. \tag{4.103}$$

The first value is used in calculations of dispersion interactions, while the second one is used in calculations of electrostatic interactions. The minimum value of the cell edge is defined as

$$r_{\text{cut}} = L / 2. \tag{4.104}$$

In the case of short-range (dispersion) interactions, $\alpha \geq 6$ (see Eq. 4.102) and the interaction potential energy $u(r)$ in Eq. 4.102 decreases very rapidly. The simplest approach consists of the neglect of the energy of dispersion interactions at $r \geq r_{\text{cut}}$:

$$u(r) = \begin{cases} u(r), r \leq r_{\text{cut}} \\ 0, r > r_{\text{cut}} \end{cases} \tag{4.105}$$

The origin of coordinates is placed at the given particle. Function 4.105 has a discontinuity at point $r \geq r_{\text{cut}}$, which can result in an instability in numerical calculations. Several approaches called *shifted and shifted-force potentials* [17] were proposed to overcome this disadvantage. Consider the first, most simple approach. In this approach, a constant is added to all pair interactions, which does not affect forces:

$$u^{\text{sh}}(r) = \begin{cases} u(r) - u(r_{\text{cut}}), r \leq r_{\text{cut}} \\ 0, r > r_{\text{cut}} \end{cases} \tag{4.106}$$

To verify the condition $r < r_{\text{cut}}$ in Eq. 4.106, it is necessary to calculate a great number of interparticle interactions, which is proportional to $\sim r^3$, because it is not known in advance which particles are contained in the sphere of radius r_{cut}, and which particles are not. In order that one would not search for particles that fall into a sphere of radius r_{cut} around the considered particle at each step of the trajectory, Verlet proposed an algorithm called the Verlet list. This algorithm is described in detail in the book [17].

In the case of interaction of charged particles ($u(r) \sim r^{-1}$) or molecules with dipole moments $u(r) \approx r^{-3}$, the problem of the cutoff of long-range electrostatic potentials arises. This problem is treated using special approaches, because it is not possible to set these interactions equal to zero at $r > r_{cut}$ by analogy with Eq. 4.106 even at very large values of cutoff radii.

Depending on the simulated system, various algorithms are used. Two approaches, the Ewald method and the reaction field method, are the most widespread ones. Both these approaches borrow ideas from electrostatics, in which it is suggested that charges contained in a spherical cavity polarize the environment. The environment acts on charges in the cavity, and the extent of its action depends on the static dielectric permittivity ε_s of the environment.

The reaction field method is widely used in the calculations of biological systems and in the simulation of ion dynamics in solutions. It is assumed that the field acting on a polar particle or an ion consists of two contributions: (i) the field creating by particles contained inside cavity V of radius r_{cut} and (ii) the reaction field of the dielectric outside the cavity arising because of the polarization of the dielectric by particles contained inside the cavity.

The reaction field E_i acting on molecule i is written as

$$E_i = \frac{2(\varepsilon_S - 1)}{2\varepsilon_S + 1}\frac{1}{r_{cut}^3}\sum_{j \in V}\mu_j \,. \tag{4.107}$$

The contribution of the reaction field to the interaction energy is $-(1/2)\mu_i E_i$.

Note that the above approach as well as the Ewald method is *phenomenological*, because it uses an empirical parameter, namely the static dielectric permittivity.

4.8 MD Realization of an *NVT* Ensemble: Analysis of the Results of Simulation

Depending on the properties under consideration, MD calculations are performed using different ensembles. For a description of properties of liquids and biological systems, the most widespread is the canonical *NVT* ensemble. It describes the statistical distribution for systems having the volume *V*, the given number of particles *N*, which can exchange energy with the environment (Section 4.2). MD calculations provide the RDF of the simulated system and the number of molecules in first and second solvate shells (see Section 4.4). The knowledge of the RDF allows one to find the average potential energy, the equation of state, and other thermodynamic parameters of the system under consideration (see Section 4.6). In particular, the second virial coefficient $B_2(T)$ (see Eq. 4.31) is written as

$$B_2(T) = -2\pi \int_0^{\infty} \left[\exp\left(-\frac{u(r)}{k_B T} \right) - 1 \right] r^2 dr . \tag{4.108}$$

The integrand decreases very rapidly, and it virtually equals zero at $r > 3\sigma$.

In addition to static properties, MD calculations allow dynamic properties, for example, the diffusion coefficient to be found. According to the Einstein equation, the diffusion coefficient D is related to the mean square of the particle displacement $\left\langle \left(\vec{r}(0) - \vec{r}(t) \right)^2 \right\rangle$ as follows:

$$\lim_{t \to \infty} \left\langle \left(\vec{r}(0) - \vec{r}(t) \right)^2 \right\rangle = 6Dt \tag{4.109}$$

In MD calculations of the diffusion coefficient by the least square method, the linear part of the graph of the function $R_2(t)$

$$R_2(t) = \left\langle \left(\vec{r}(0) - \vec{r}(t) \right)^2 \right\rangle = 6Dt . \tag{4.110}$$

4.9 Method of Time Correlation Functions

One of the indicators giving averaged characteristics of the motion of particles is the time correlation function of velocity. The investigation of this quantity by the MD method gave answers to a great number of questions necessary for understanding the main regularities of liquid particles:

- How are collective motions performed in the system?

- What is the role of spatial and time correlations in the motion of particles, and how do these correlate with data obtained from stochastic kinetic equations?

Let us introduce the notion of time correlation functions in the classical limit. Let $p(t)$ and $q(t)$ be momenta and coordinates required to specify the system, while $p = p(0)$ and $q = q(0)$ are momenta and coordinates at the initial instant of time $t = 0$. Then, $p(t)$ and $q(t)$ are given by the equations

$$p(t) = p(p,q;t)$$

and

$$q(t) = q(p,q;t). \tag{4.111}$$

Suppose $A\{p(t),q(t)\}$ is a certain function in the phase space. Using Eq. 4.111, we can write

$$A\{p(t),q(t)\} = A(p,q,t) = A(t). \tag{4.112}$$

Let us define the classical time correlation function $C(t)$ as

$$C(t) = \langle A(0)A(t)\rangle = \int\int dqdpA(p,q;0)A(p,q;t)f(p,q). \tag{4.113}$$

Here, $f(p,q)$ is the equilibrium distribution function, and $dpdq \equiv dp_1 \ldots dp_N dq_1 \ldots dq_N$. If $\mathbf{A}(t)$ is a vector function, Eq. 4.113 is written in the form

$$C(t) = \langle \mathbf{A}(0)\mathbf{A}(t)\rangle = \int\int dqdp\mathbf{A}(p,q;0) \bullet \mathbf{A}(p,q;t)f(p,q). \tag{4.114}$$

As an example, let us consider the case when $\mathbf{A}(t)$ is the velocity of a single molecule. Then, Eq. 4.114 can be written in the form

$$C(t) = \langle \vec{v}(0)\vec{v}(t)\rangle. \tag{4.115}$$

Note that, for the calculation of $C(t)$ in the simplest case, it is necessary to solve the equation of motion of a particle surrounded by other particles and to average the initial coordinates over an equilibrium ensemble. Because $\vec{v}(t)$ depends on momenta and coordinates of other particles of the system, the exact solution of the problem cannot be easily found.

An attractive feature of the considered formalism is the fact that the final formulas for transfer coefficients are of rather general character and are independent of nuances of the model used and are not connected with the system density. For example, the diffusion coefficient (see Section 4.3.1) is presented in the form

$$D = \frac{1}{3}\int_{0}^{\infty}\langle\vec{v}(0)\vec{v}(t)\rangle dt. \quad (4.116)$$

This expression does not depend on density and is applicable in the case of forces depending on angles, which arise in the description of polyatomic molecules.

The infrared (IR) spectrum of the simulated system can be obtained by the Fourier transformation of the autocorrelator of the dipole moment $\vec{M}(t)$ calculated at each step of the trajectory

$$I(\omega) = \omega\left(\frac{2}{\varepsilon_0 \mathbf{c}\hbar n}\right) th\left(\frac{\hbar\omega}{2k_B T}\right) \mathrm{Re}\int_{0}^{\infty} e^{i\omega t}\langle\vec{M}(t)\vec{M}(0)\rangle dt. \quad (4.117)$$

Here, $I(\omega)$is the relative absorption intensity at frequency ω, **c** is the velocity of light, ε_0 is the vacuum permittivity, and n is the optical permittivity, which is assumed to be constant.

The IR spectrum of water calculated by Eq. 4.117 with the dipole moment calculated by the MD method with classical force fields was reported in Ref. [18]. In the cited work, the *NVT* ensemble was used, the step was 1 *fs*, and the cutoff radius was 1.4 nm. Electrostatic interactions were described by the reaction field method. The length of the trajectory was 600 ps, including 100 ps for thermostating.

Appendix: Software for MD Simulation Used in Calculations of Liquids and Disordered Condensed Systems

(1) VASP (Viena Ab initio Simulation Package). Program package for calculations by DFT methods in the plane-wave basis set with periodic boundary conditions. Payware.

(2) CPMD (Car–Parrinello Molecular Dynamics). Program package for calculations by DFT methods (generalized gradient approximations) in the plane-wave basis set with periodic boundary conditions. CPMD is specially developed for parallelized MD calculations. As distinct from VASP. Freeware.

(3) AMBER (Assisted Model Building with Energy Refinement). This program is used for proteins, nucleic acids, and other classes of molecules. Payware.

(4) CHARMm (Chemistry at HARvard Macromolecular mechanics). The program is designed for calculations of various systems, from small molecules to solvated complexes of biological macromolecules. Payware.

(5) GROMACS (Groningen Simulation Package). Program package for calculations of biological macromolecules in aqueous solutions. This software in its characteristics is close to AMBER and CHARMm. Freeware.

4.11 Conclusion

In this chapter, an attempt is made to briefly present elements of statistical mechanics as applied to classical nonideal systems and describe the main computational methods used in computer simulations of micro- and macroscopic properties of disordered condensed systems.

Note that the fast development of computational technique and the corresponding software makes possible calculations of various properties of condensed systems without using a mechanical model of molecules and the approximation of pair potentials (classical force fields). We mention first CPMD, in which interactions in the system are calculated by DFT methods.

An important achievement of recent years is the widespread use of the technique of time correlation functions in computer simulation. In the long term, this will provide the possibility of describing not only equilibrium properties but also nonequilibrium characteristics. This problem can be solved in near future.

4.12 Further Reading

1. M. Tuckerman, *Statistical Mechanics. Theory and Molecular Simulations*, Oxford University Press, New York, 2010, 713 p.
2. D. Marx and J. Hutter, *Ab Initio Molecular Dynamics: Basic Theory and Advanced Methods*, Cambridge University Press, Cambridge, 2009, 567 p.

3. N. G. Van Kampen, *Stochastic Processes in Physics and Chemistry*, North Holland, Amsterdam, 2007, 464 p.
4. D. A. McQuarrie, *Statistical Mechanics*, University Science Books, Sausalito, CA, 2000, 641 p.

References

1. L. D. Landau and E. M. Lifshitz, *Statistical Physics*, 3d ed., *Course of Theoretical Physics, Part 1*, Vol. 5, Translated from Russian by J. B. Sykes and M. J. Kearsley, Oxford, England Pergamon Press, 563 p.
2. I. G. Kaplan, *Intermolecular Interactions: Physical Picture, Computational Methods and Model Potentials*, Wiley, Chichester, 2006, 367 p.
3. N. N. Bogoliubov, *On Some Statistical Methods in Mathematical Physics*, Academy of Sciences of Ukraine, Kiev, 1945 (in Russian).
4. N. N. Bogoliubov, *Problems of Dynamic Theory in Statistical Physics*, Technical Information Service, Oak Ridge, Tennessee, 1960.
5. B. J. Alder and T. E. Wainwright, Phase transition for a hard sphere system, *J. Chem. Phys.*, 1957, **27**, 1208.
6. W. W. Wood and J. D. Jacobson, Preliminary results from a recalculation of the Monte Carlo equation of state of hard spheres, *J. Chem. Phys.*, 1957, **27**, 1207.
7. M. Tuckerman, K. Laasonen, M. Sprik, and M. Parrinello, Ab initio molecular dynamics simulation of the solvation and transport of hydronium and hydroxyl ions in water, *J. Chem. Phys.*, 1995, **103**, 150–161.
8. A. N. Morozov and S. H. Lin, Modeling of folding and unfolding mechanisms in alanine-based r-helical polypeptides, *J. Phys. Chem. B*, 2006, **110**, 20555.
9. D. Riccardi, P. Schaefer, Y. Yang, H. Yu, N. Ghosh, X. Prat-Resina, P. König, G. Li, D. Xu, H. Guo, M. Elstner, and Q. Cui, Development of effective quantum mechanical/molecular mechanical (QM/MM) methods for complex biological processes, *J. Phys. Chem. B*, 2006, **110**, 6458.
10. N. Sieffert, M. Bühl, M.-P. Gaigeot, and C. A. Morrison, Liquid methanol from DFT and DFT/MM molecular dynamics simulations, *J. Chem. Theory Comput.*, 2013, **9**, 106.
11. G. G. Maisuradze, P. Senet, C. Czaplewski, A. Liwo, and H. A. Scheraga, Investigation of protein folding by coarse-grained molecular dynamics with the UNRES force field, *J. Phys. Chem. A*, 2010, **114**, 4471.

12. K. S. Schweizer and J. G. Curro, PRISM theory of the structure, thermodynamics, and phase transitions of polymer liquids and alloys, *Adv. Polym. Sci.*, 1994, **116**, 319.

13. K. S. Schweizer and J. G. Curro, Integral equation theories of the structure, thermodynamics, and phase transitions of polymer fluids, *Adv. Chem. Phys.*, 1997, **98**, 1.

14. Y. Yonetani, Y. Maruyama, F. Hirata, and H. Kono, Comparison of DNA hydration patterns obtained using two distinct computational methods, molecular dynamics simulation and three-dimensional reference interaction site model theory, *J. Chem. Phys.*, 2008, **128**, 185102.

15. L. Verlet, Computer «experiments» on classical fluids. I. Thermodynamical properties of Lennard-Jones molecules, *Phys. Rev.*, 1967, **159**, 98.

16. N. Metropolis, A. W. Rosenbluth, M. N. Rosenbluth, A. H. Teller, and E. Teller, Equation of state calculations by fast computing machines, *J. Chem. Phys.*, 1953, **21**, 1087.

17. M. P. Allen and D. J. Tildesley, *Computer Simulation of Liquids*, Oxford University Press, New York, 1987, 385 p.

18. M. Praprotnik, D. Janežič, and J. Mavri, Temperature dependence of water vibrational spectrum: a molecular dynamics simulation study, *J. Phys. Chem. A*, 2004, **108**, 11056.

Chapter 5

Vibronic Interactions

5.1 Optical Transitions

The emission intensity $S(\Omega)$ for an electronic transition between the upper state b and the lower state a of a single emitter in an isotropic medium can be written as

$$S(\Omega)=\frac{4\Omega^4 n}{3\mathbf{c}^3}I_{ab}(\Omega),$$

while the absorption cross section is

$$\sigma(\Omega)=\frac{4\pi^2\Omega}{3\mathbf{c}n}I_{ba}(\Omega).$$

Here, Ω is the emitted/absorbed photon frequency, n is the refractive index of the medium, and **c** is the velocity of light.[a] The quantity $I_{ab}(\Omega)$ is called the line shape function and is defined as a thermally averaged sum over all individual electron vibrational (vibronic) transitions

$$I_{ab}(\Omega)=\sum_{p,q}P_{bq}\left|\langle\chi_{ap}\,|\,M_{ab}(Q)\,|\,\chi_{bq}\rangle\right|^2\delta\left(\left(E_{bq}-E_{ap}\right)/\hbar-\Omega\right).\quad(5.1)$$

[a]For processes in dielectric media, the local field factor ($E_{\text{medium}}=E_{\text{vacuum}}$) is generally taken into account, for example, in the Lorentzian form $l=(n^2+2)/3$.

Multiscale Modeling in Nanophotonics: Materials and Simulation
Alexander Bagaturyants and Mikhail Vener

ISBN 978-981-4774-40-6 (Hardcover), 978-1-315-10976-3 (eBook)
www.panstanford.com

Here $M_{ab}(Q)=\langle\Phi_a(r,Q)|M|\Phi_b(r,Q)\rangle_r$ is an electronic transition moment (electric dipole, magnetic dipole, etc.); $\Phi_{a,b}$ are electronic wave functions of the lower (a) and upper (b) states; χ_{ap}, χ_{bq} are vibrational functions; and E_{ap}, E_{bq} are vibronic energy levels in the corresponding electronic states.

$$P_{bq}=\exp(-\beta E_{bq})\Big/\sum_r\exp(-\beta E_{br}) \tag{5.2}$$

are Boltzmann weights of vibrational levels in the initial electronic state, and, $\beta=1/k_BT$, where k_B is the Boltzmann constant and T is absolute temperature. In the case of small amplitudes of vibrational displacements $Q_j - Q_{j0}$, , the electronic transition moment function can be expanded into a power series in $Q_j - Q_{j0}$

$$M_{ab}(Q)=M_{ab}(Q_0)+\sum_j\left.\frac{\partial M_{ab}(Q)}{\partial Q_j}\right|_{Q=Q_0}(Q_j-Q_{j0})+\ldots$$

Here, Q_0 is a certain initial nuclear configuration of the system.[b] Assume that this expansion can be restricted to the first term (so-called *Condon approximation*). Then

$$\left|\langle\chi_{ap}|M_{ab}(Q)|\chi_{bq}\rangle\right|^2\approx\left|M_{ab}(Q_0)\right|^2\left|\langle\chi_{ap}|\chi_{bq}\rangle\right|^2 \tag{5.3}$$

The quantity $\left|\langle\chi_{ap}|\chi_{bq}\rangle\right|^2$ is called the Franck–Condon (FC) factor. In the Condon approximation, the line shape function (Eq. 5.1) is transformed to

$$I_{ab}(\Omega)=\left|M_{ab}(Q_0)\right|^2\sum_{i,j}P_{bj}\left|\langle\chi_{ap}|\chi_{bq}\rangle\right|^2\delta\left((E_{bq}-E_{ap})/\hbar-\Omega\right). \tag{5.4}$$

If only relative rather than integral intensities of vibronic lines are of interest, it is reasonable to introduce the so-called normalized spectrum

$$G_{ab}(\Omega)=\sum_{i,j}P_{bj}\left|\langle\chi_{ai}|\chi_{bj}\rangle\right|^2\delta\left((E_{bq}-E_{ap})/\hbar-\Omega\right) \tag{5.5}$$

---

[b]"Initial" here means "reference," that is, not obligatory "starting." The choice of the initial configuration is, strictly speaking, ambiguous; see footnote *e* later.

$$\int_{-\infty}^{+\infty} G_{ab}(\Omega)d\Omega = 1$$

and then to calculate the $G_{ab}(\Omega)$ function or some approximation to it.

5.2 The Lax Model

In the case of electron vibrational transitions in large molecules, clusters, and condensed media, lines responsible for individual electron vibrational transitions overlap each other so extensively that the entire spectrum looks as their envelop, sometimes with a discernible vibronic structure. Therefore, a direct calculation of individual FC factors and their direct summation in Eqs. 5.4 and 5.5 often becomes an unreasonably cumbersome task.

In his fundamental work, Lax [1] expressed the normalized spectrum in the Condon approximation (Eq. 5.5) as a Fourier integral over time t

$$G_{ab}(\Omega) = \frac{1}{2\pi}\int_{-\infty}^{+\infty} \exp(-i\Omega t)G_{ab}(t)dt \tag{5.6}$$

Here, $G_{ab}(t)$ is a generating function (GF) of the normalized spectrum. Formally, Eq. 5.6 can be obtained if the integral representation of the δ function is used in Eq. 5.5:

$$\delta\left(\left(E_{bq} - E_{ap}\right)/\hbar - \Omega\right) = \frac{1}{2\pi}\int_{-\infty}^{+\infty} \exp\left[i\left(\left(E_{bq} - E_{ap}\right)/\hbar - \Omega\right)t\right]dt$$

$$\Rightarrow \quad G_{ab}(\Omega) = \frac{1}{2\pi}\int_{-\infty}^{+\infty}\left[\sum_{p,q} P_{bq}\left|\left\langle \chi_{ap} \mid \chi_{bq}\right\rangle\right|^2 \exp\left(iE_{bq}t/\hbar\right)\exp\left(-iE_{ap}t/\hbar\right)\right]$$

$$\exp(-i\Omega t)dt;$$

$$G_{ab}(t) = \frac{\sum_{p,q}\left|\left\langle \chi_{ap} \mid \chi_{bq}\right\rangle\right|^2 \exp\left(\left(it/\hbar - \beta\right)E_{bq}\right)\exp\left(-iE_{ap}t/\hbar\right)}{\sum_r \exp\left(-\beta E_{br}\right)}.$$

After some transformations, the GF can be written as

$$G_{ab}(t)=\frac{\mathrm{Tr}\left[\exp\left(-iH_a t/\hbar\right)\exp\left(H_b\left(it/\hbar-\beta\right)\right)\right]}{\mathrm{Tr}\left[\exp\left(-\beta H_b\right)\right]}.$$

Here, H_b and H_a are vibrational Hamiltonians for the upper and lower electronic states, respectively, and Tr is the symbol of an operator trace.

In some cases, $G_{ab}(t)$ can be calculated in a closed form, for example, within the following simple harmonic model with a linear electron-phonon coupling.[c] Suppose that the adiabatic potentials of electronic states are described by harmonic surfaces of the same shape and differ only in positions of equilibrium (Fig. 5.1):

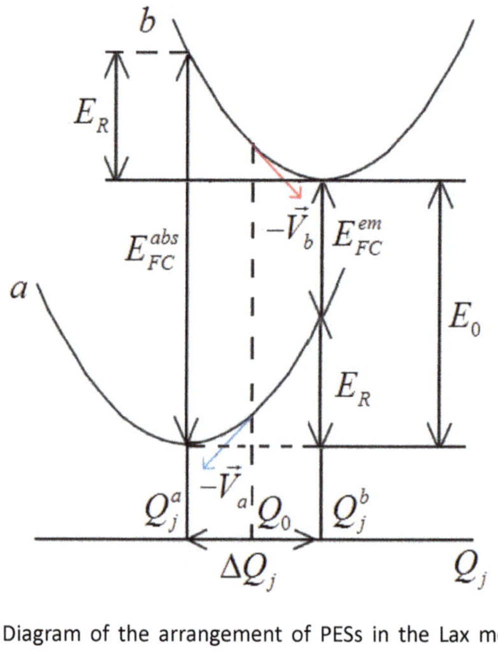

Figure 5.1 Diagram of the arrangement of PESs in the Lax model. $E_{\mathrm{FC}}^{\mathrm{abs}}$ is the vertical transition energy for absorption, $E_{\mathrm{FC}}^{\mathrm{em}}$ is the same for emission, E_0 is the energy of the purely electronic (phononless) transition, and E_{R} is the reorganization energy [12].

[c]This model is one of most popular ones in the literature, starting from the middle of the twentieth century and up to now; a general solution of the problem of the shape of an optical spectrum within this model was first given by Lax [1].

$$E^a(Q) = E^a\left(Q^0\right) + \sum_j V_j^a\left(Q_j - Q_j^0\right) + \frac{1}{2}\sum_j \omega_j^2\left(Q_j - Q_j^0\right)^2$$

$$E^b(Q) = E^b\left(Q^0\right) + \sum_j V_j^b\left(Q_j - Q_j^0\right) + \frac{1}{2}\sum_j \omega_j^2\left(Q_j - Q_j^0\right)^2 \quad (5.7)$$

Here, $Q = \{Q_j\}$ are harmonic normal modes in the neighborhood of the selected reference configuration Q_0; ω_j are the corresponding harmonic frequencies; and $V_j^{a,b} = \partial E^{a,b}/\partial Q_j\big|_{Q=Q_0}$ are projections of gradients of potential energy surfaces (PESs) along the normal modes in the reference configuration Q_0[d]. From Eq. 5.7, it follows that

$$\partial E^{a,b}(Q)/\partial Q_j = V_j^{a,b} + \omega_j^2\left(Q_j - Q_j^0\right);$$
$$\partial E^b(Q)/\partial Q_j - \partial E^a(Q)/\partial Q_j \equiv V_j^b - V_j^a = \text{const}, \quad (5.8)$$

that is, for harmonic surfaces of type Eq. 5.2, the difference of the projections of their gradients is actually independent of the choice of the reference configuration Q_0, which, thereby, does not enter into all the subsequent equations (Eqs. 5.9–5.12).[d]

The relative equilibrium positions of the lower and upper electronic states, $Q_j^a - Q_j^0$ and $Q_j^b - Q_j^0$, are expressed as

$$Q_j^{a,b} - Q_j^0 = -V_j^{a,b}/\omega_j^2$$

In fact, this is a well-known formula of the Newton–Raphson algorithm for a function minimum search; in the case of harmonic surfaces, it gives an exact answer. From this, the displacement of equilibrium positions of two states in normal coordinates ΔQ can be presented as

$$\Delta Q_j = Q_j^b - Q_j^a = -\left(V_j^b - V_j^a\right)/\omega_j^2 .$$

These points are discussed in detail in the dissertation by M. K. Kretov [12]. Now, consider another important quantity, namely the dimensionless Huang–Rhys (HR) factor

$$S_j = \omega_j \Delta Q_j^2/2\hbar = \left(V_j^b - V_j^a\right)^2/2\hbar\omega_j^3 . \quad (5.9)$$

[d]If, however, the discussed parameters of the spectral model are estimated from realistic PESs, these become dependent on the choice of the initial configuration.

The physical sense of the HR factor for mode j is the average number of vibrational quanta that are created on this mode upon an electron vibrational transition. This becomes evident if the reorganization energy E_R (marked in Fig. 5.1) is expressed through HR factors

$$E_R = \sum_j S_j \hbar\omega_j \,. \tag{5.10}$$

It follows from Eq. 5.8 that the displacement of equilibrium positions and the HR factors are independent of the choice of the initial configuration Q_0. An explicit expression for the GF, $G_{ab}(t)$, of the normalized emission spectrum was derived within the above model of PESs by Lax [1]:

$$\begin{aligned} G_{ab}(t) &= \exp\left(i\Omega_0 t - \sum_j S_j \left[(2\bar{n}_j + 1)(1 - \cos\omega_j t) + i \sin\omega_j t \right] \right) \\ &= \exp\left(i\Omega_0 t + f(t) \right) \end{aligned} \tag{5.11}$$

Here, S_j are HR factors (Eq. 5.9), $\Omega_0 = E_0/\hbar$ is the frequency of the purely electronic (phononless) transition (Fig. 5.1), and $\bar{n}_j = \left[\exp\left(\hbar\omega_j / k_B T \right) - 1 \right]^{-1}$ are Planck's thermal populations of vibrational modes. For the absorption spectrum, the Lax formula takes a similar form

$$\begin{aligned} G_{ba}(t) &= \exp\left(i\Omega_0 t + f^*(t) \right), \\ f^*(t) &= f(-t) \end{aligned} \tag{5.12}$$

with the same microscopic parameters as in Eq. 5.7.

In summary, note that the theoretical result under discussion has been derived by Lax under the assumption that the normal modes Q_j and harmonic frequencies ω_j in both the considered electronic states are identical (see Eq. 5.7) and that the Condon approximation (Eq. 5.3) is applicable. The most general formal expression for the GF of an optical spectrum within the model of harmonic surfaces and also its further generalization to the case of non-Condon transitions have been later obtained by Kubo and Toyazawa [1].

5.3 Pekar's Model

Pekar in his seminal review [2] presented several particular cases when the infinite series of FC factors (Eq. 5.4) for an optical transition band could be approximately summed within the Condon approximation. One of Pekar's results based on the same multidimensional harmonic model as in the Lax approach, but with an additional assumption of large heat release in the electronic transition, will be considered next. This assumption usually means [3] a large reorganization energy (Eq. 5.5), that is, the case of strong electron-phonon coupling ($S_j >> 1$).

An original derivation of Pekar's equation is given in the work by Pekar and Krivoglaz [4] and employs a complicated technique of a theory of functions of a complex variable. However, the same equation can be obtained in a much simpler way if the Lax approach described earlier is used [1, 5] in the so-called short time limit [6–6].[e] In this approach, the exponent in the GF, $G_{ab}(t)$, for an emission spectrum (Eq. 5.6) is expanded into a series up to terms quadratic in t

$$G_{ab}(t) = \exp\left\{ i\Omega_0 t - \sum_j S_j\left[\left(2\bar{n}_j+1\right)\left(1-\cos\omega_j t\right)+i\sin\omega_j t\right]\right\}$$

$$= \exp\left\{ i\left(\Omega_0 - E_\mathrm{R}/\hbar\right)t - g'' t^2/2 + o\left(t^3\right)\right\},$$

where $$E_\mathrm{R} = \sum_j S_j \hbar\omega_j \,,\; g'' = \sum_j S_j \omega_j^2\left(2\bar{n}_j+1\right),$$

and $$\bar{n}_j = \left[\exp\left(\hbar\omega_j/k_\mathrm{B}T\right)-1\right]^{-1}.$$

After taking the Fourier transform (Eq. 5.6), Pekar's formula is obtained,

$$G_{ab}(\Omega) = \frac{1}{\sqrt{2\pi g''}}\exp\left(-\frac{\left(\Omega_\mathrm{FC}^\mathrm{em}-\Omega\right)^2}{2g''}\right), \qquad (5.13)$$

[e]Physical interpretation of the short time limit within the wave packet formalism will be given in the next section.

where

$$g'' = \sum_j S_j \omega_j^2 (2\bar{n}_j + 1) = \sum_j S_j \omega_j^2 \cot h\left(\frac{\hbar\omega_j}{2kT}\right) \quad (5.14)$$

and $\Omega_{\rm FC}^{\rm em} = \Omega_0 - E_{\rm R}/\hbar = E_{\rm FC}^{\rm em}/\hbar$ is the frequency of the vertical (FC) transition for emission (see Fig. 5.1). For the absorption spectrum, a similar formula is obtained starting from Eq. 5.12,

$$G_{ba}(\Omega) = \frac{1}{\sqrt{2\pi g''}} \exp\left(-\frac{\left(\Omega_{\rm FC}^{\rm abs} - \Omega\right)^2}{2g''}\right), \quad (5.15)$$

where $\Omega_{\rm FC}^{\rm abs} = \Omega_0 + E_{\rm R}/\hbar = E_{\rm FC}^{\rm abs}/\hbar$.

Thus, in the limit of a strong electron-phonon coupling, that is, a large heat release, Pekar's model predicts a structureless band of Gaussian shape with a maximum at the frequency of the vertical (FC) transition. In this case, the spectral band, apart from the maximum position, is characterized by only one parameter, the full width at half maximum (FWHM) $\delta\omega$, which is expressed for a Gaussian band through the dispersion parameter g'' (Eq. 5.14) as follows [3]:

$$\delta\omega = 2\sqrt{2\ln 2}\sqrt{g''} \quad (5.16)$$

In this case, the absorption and emission bands coincide in shape and width, and their maxima are located symmetrically ($\Omega_0 \pm E_{\rm R}/\hbar$) with respect to the phononless transition frequency $\Omega = \Omega_0$ (*mirror symmetry law*).

Note in addition that a cubic term is also present in the exponent of the original Pekar's formula from Ref. [5]:

$$G(\Omega) = \frac{1}{\sqrt{2\pi g''}} \exp\left(-\frac{\left(\Omega - \Omega_{\rm FC}\right)^2}{2g''} \mp \frac{g'''\left(\Omega - \Omega_{\rm FC}\right)^3}{6\left(g''\right)^2} + \ldots\right) \quad (5.17)$$

The minus or plus sign corresponds to the emission or absorption case, respectively. The cubic term in Pekar's formula is commonly associated with the third moment, that is, with the asymmetry of the spectral band [4]. However, as is noted in Pekar's review [3], the effect of this term in the case of a large heat release is insignificant, at least around the band maximum.

From a practical point of view, the spectral band shape in Pekar's model is determined by the same microscopic parameters as in the Lax model.

5.4 Relaxation of the Nuclear Subsystem after an Electronic Transition

The GF formalism, introduced initially by Lax only as a formal technique facilitating the calculation of the FC profile of a vibronic band, turned out to be an exceptionally useful concept because it allows the problem of calculating the FC profile of a vibronic band to be connected with the dynamics of the nuclear subsystem after an electronic transition. According to the classical FC principle, the nuclei immediately after an electronic transition are characterized by the same positions $Q = \{Q_j\}$ and momenta $P = \{P_j\}$ as before the transition. In quantum mechanics, these correspond to expectation values of coordinates and momenta, $\langle Q_j \rangle$ and $\langle P_j \rangle$. The time evolution of these quantities is determined by the Ehrenfest quantum equations, which, for independent harmonic oscillators, take the form

$$\langle \dot{Q}_j \rangle_t = \langle P_j \rangle_t; \qquad \langle \dot{P}_j \rangle_t = -\omega_j^2 \left[\langle Q_j \rangle_t - Q_j^a \right], \tag{5.18}$$

while the forces acting on nuclei correspond to the derivatives of the adiabatic potential of the final electronic state (lower one, a, for emission; see Fig. 5.1). Here and later, as well as in the Lax model (Section 5.2), mass-weighted normal coordinates (the same in both the electronic states) are considered. If temperature distributions over initial vibrational levels are not taken into account ($T \to 0$, Planck's populations $\bar{n}_j \to 0$), the initial conditions $\langle Q_j \rangle_0$ and $\langle P_j \rangle_0$ correspond to the ground vibrational state of the initial electronic term (in the case of emission, the upper one, b; see Fig. 5.2)

$$\chi_{b0}(Q) = \prod_j \left(\omega_j / \pi \right)^{1/4} \exp\left[-\omega_j \left(Q_j - Q_j^b \right)^2 / 2 \right] \tag{5.19}$$

$$\langle Q_j \rangle_0 = Q_j^b, \qquad \langle P_j \rangle_0 = 0.$$

A partial solution of Eq. 5.18 corresponding to initial conditions in Eq. 5.19, that is, the quantum law of motion, takes the form

$$\langle Q_j \rangle_t = Q_j^a + \left(Q_j^b - Q_j^a\right)\cos\omega_j t;\ \langle P_j \rangle_t$$
$$= \langle \dot{Q}_j \rangle_t = -\,\omega_j\left(Q_j^b - Q_j^a\right)\sin\omega_j t, \tag{5.20}$$

which is formally the same as in classical mechanics: the expectation value of each normal coordinate $\langle Q_j \rangle$ executes harmonic vibrations in the neighborhood of a new equilibrium position Q_j^a on the PES of the final (lower, a, for emission; Fig. 5.2) electronic term with an amplitude $\left|Q_j^b - Q_j^a\right|$ and cyclic frequency ω_j. The initial wave function of the form Eq. 5.20 does not correspond, generally speaking, to a stationary state on the new PES and evolves in the subsequent instances of time according to the following equation [6]:

$$\chi(Q,t) = \prod_j \left(\omega_j/\pi\right)^{1/4}$$
$$\exp\left[-(1/2)\omega_j\left(Q_j - \langle Q_j \rangle_t\right)^2 + i\langle P_j \rangle_t Q_j + i\gamma_t\right], \tag{5.21}$$

where γ_t is a variable phase, insignificant for the subsequent analysis. A wave function given in Eq. 5.21 describes a *coherent* state [7] of a multidimensional harmonic oscillator, or, in other words, a Gaussian wave packet moving along a classical trajectory (Eq. 5.20); see Fig. 5.2.

An important relation between the wave packet dynamics described earlier and the GF of an optical spectrum was ascertained by Heller et al. [6, 10]. If the temperature distribution over initial vibrational levels is neglected, the GF (Eq. 5.11) up to a phase factor is reduced to the autocorrelation function (AF) of the wave packet (Eq. 5.21)

$$G_{ab}(t) \xrightarrow[T\to 0]{} e^{i\Omega_0 t}\langle \chi(Q,t=0) | \chi(Q,t) \rangle$$
$$= e^{i\Omega_0 t}\prod_j \exp\left(-S_j\left[1 - e^{-i\omega_j t}\right]\right), \tag{5.22}$$

which provides a pictorial interpretation of the variation of the GF with time based on the form wave function (Eq. 5.21) and classical trajectories (Eq. 5.22). The AF $\langle \chi(Q,t=0) | \chi(Q,t) \rangle$ is, in essence, an overlap integral between the wave packet (Eq. 5.21) in the initial and subsequent instants of time, which consists of a product of overlap integrals over all vibrational modes.

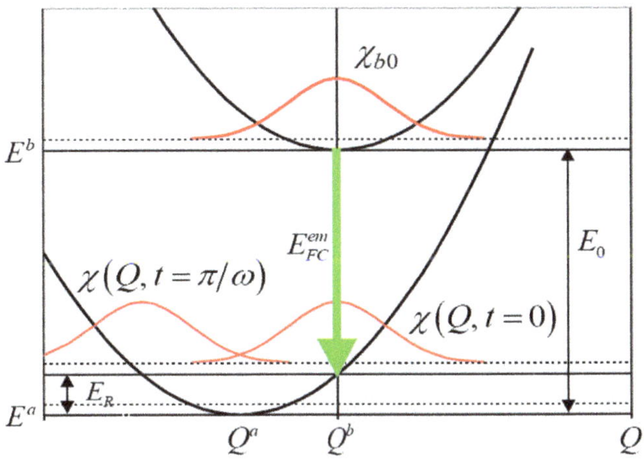

Figure 5.2 Schematic view of the evolution of a quantum wave packet (coherent state) on a harmonic PES of the lower term. Q^a, Q^b are minimum configurations, and E^a, E^b are the corresponding minimum energies; dashed lines are the same, taking into account zero-point vibrations; E_{FC}^{em} is the vertical emission energy; E_0 is the energy of the purely electronic (phononless) transition; and E_R is the reorganization energy. The other designations are explained in the dissertation [12].

At the initial instant of time ($t = 0$), each of the multipliers in the AF is, evidently, equal to 1 in magnitude (maximum overlap). In the subsequent instances of time, it either remains equal to 1 in magnitude (for those modes along which the equilibrium position is not displaced upon the transition, that is, the HR factor S_j, Eq. 2.4, equals zero) or varies periodically with a frequency of the corresponding oscillator ω_j because the probability density maximum of the wave packet (Eq. 5.21) moves according to classical law (Eq. 5.20). The AF (Eq. 5.22), as a whole, also has an oscillating character, but not periodical in the general case, because different normal coordinates do not, in general, simultaneously return to their initial values Q_j^b, corresponding to the geometry of the FC transition when the overlap with the initial state becomes again maximum in magnitude [6]. Hence, the repeated bursts of the multidimensional overlap integral are, in general, systematically suppressed, which is associated with a phase mismatch of particular vibrations.

A qualitative illustration of the possible behavior of the magnitude of the AF of a wave packet and the possible behavior of the corresponding vibronic spectrum in given in Ref. [10]. It is possible to distinguish several (three, in this case) timescales of AF variations, which correspond to specific characteristic structures in the frequency spectrum.

By virtue of the uncertainty principle (which, in this case, is only a physical consequence of the mathematical properties of the Fourier transform connecting the spectrum and the GF according to Eq. 5.6), the narrowest time intervals correspond to the most diffuse structures in the frequency spectrum and vice versa. Thus, a rapid initial decline of the AF in magnitude (which corresponds to the initial removal of the wave packet from the FC region in the period of time of order T_1) corresponds to the half width of the envelop of the entire vibronic band ($\sim 1/T_1$). If subsequently the wave packet returns at least into an immediate neighborhood of the initial (FC) geometry, which appears as characteristic spikes of the AF magnitude after equal time intervals T_2, a vibronic structure arises in the spectrum with a spacing of $2\pi/T_2$. Finally, the systematic decrease of the AF magnitude at longest times ($\sim T_3$) in this example corresponds to the broadening of vibronic peaks of order $1/T_3$.

The above example is simplified, but it illustrates various particular situations. For example, from the viewpoint of wave packet dynamics, the disappearance of the vibronic structure of a band in the case of strong electron-phonon coupling ($S_j >> 1$) can be interpreted as rapid systematic decay of the GF (virtually, already in a time of order of the initial decay T_1) and the absence of recurring oscillations. As a result, the spectrum will consist only of a structureless envelop. On the other hand, the restriction by short times in the derivation of Pekar's formula (Eqs. 5.15–5.17) is, in some sense, analogous to breaking the Fourier transform of the GF after its initial decline $\sim T_1$, which leads to a Gaussian envelop of the FC band profile. An analysis of the subsequent evolution of the system allows, at least in principle, details of the vibronic structure in the band to be revealed. The wave packet formalism, actively developed in the 1970s–1980s by Heller et al. [6, 8], additionally allows for a numerical investigation of the evolution of the wave packet on an anharmonic surface, where new effects and related problems arise, such as spreading of the wave packet because of variation of the local curvature of the anharmonic surface along the classical trajectory and also gradual deviation of its shape from the initial Gaussian (Eq. 5.21), which appears at long times of evolution [9].

Practical aspects of calculations of the vibronic structure of an optical band are considered, for example, in Refs. [7, 10–14]; see also Refs. [15, 16].

References

1. M. Lax, The Franck–Condon principle and its application to crystals, *J. Chem. Phys.*, 1952, **20**, 1752–1760.
2. R. Kubo and Y. Toyozawa, Application of the method of generating function to radiative and non-radiative transitions of a trapped electron in a crystal, *Prog. Theor. Phys.*, 1955, **13**, 160.
3. S. I. Pekar, On the effect of lattice deformation by electrons on the optical and electrical properties of crystals, *Usp. Fiz. Nauk*, 1953, **50**, 197–252 (English transl.: *Sov. Phys.-Usp.*).
4. Yu. E. Perlin, Modern methods of the theory of multiphonon processes, *Usp. Fiz. Nauk*, 1963, **80**, 553–595 (English transl.: *Sov. Phys.-Usp.*).
5. M. A. Krivoglaz and S. I. Pekar, The shape of light absorption and luminescence of impurities in dielectrics, *Trudy Instituta Fiziki AN USSR*, 1953, 37–70.
6. D. J. Tannor and E. J. Heller, Polyatomic Raman scattering for general harmonic potentials, *J. Chem. Phys.*, 1982, **77**, 202–218.
7. F. Neese, T. Petrenko, D. Ganyushin, and G. Olbrich, Advanced aspects of ab initio theoretical optical spectroscopy of transition metal complexes: multiplets, spin-orbit coupling and resonance Raman intensities, *Coord. Chem. Rev.*, 2007, **251**, 288–327.
8. C. K. Chan and J. B. Page, Temperature effects in the time correlator theory of resonance Raman scattering, *J. Chem. Phys.*, 1983, **79**, 5234–5250.
9. D. Landau and E. M. Lifshitz, *Quantum Mechanics: Non-Relativistic Theory*, 3rd ed., *Course of Theoretical Physics*, Vol. 3, Pergamon, Oxford, 1977.

10. E. J. Heller, The semiclassical way to molecular spectroscopy, *Acc. Chem. Res.*, 1981, **14**, 368–375.

11. E. J. Heller, Guided Gaussian wave packets, *Acc. Chem. Res.*, 2006, **39**(2), 127–134.

12. M. K. Kretov, *Quantum-Chemical Study of Vibronic Bands of Manganese (II) Ions in Inorganic Matrices*, PhD thesis, M. V. Lomonosov Moscow State University, Russia, 2012.

13. P. V. Yurenev, M. K. Kretov, A. V. Scherbinin, and N. F. Stepanov, Environmental broadening of the CTTS bands: the hexaammineruthenium(II) complex in aqueous solution, *J. Phys. Chem. A*, 2010, **114**, 12804.

14. M. K. Kretov, I. M. Iskandarova, B. V. Potapkin, A. V. Scherbinin, A. M. Srivastava, and N. F. Stepanov, Simulation of structured ^{4}T1-6A1 emission bands of Mn^{2+} impurity in Zn_2SiO_4: a first-principle methodology, *J. Lumin.*, 2012, **132**, 2143–2150.

15. M. K. Kretov, A. V. Scherbinin, and N. F. Stepanov, Simulating the structureless emission bands of Mn^{2+} ions in $ZnCO_3$ and $CaCO_3$ matrices by means of quantum chemistry, *Russ. J. Phys. Chem. A*, 2013, **87**(2), 245–251.

16. P. S. Rukin, A. Ya. Freidzon, A. V. Scherbinin, V. A. Sazhnikov, A. A. Bagaturyants, and M. V. Alfimov, Vibronic bandshape of the absorption spectra of dibenzoylmethanatoboron difluoride derivatives: analysis based on ab initio calculations, *Phys. Chem. Chem. Phys.*, 2015, **17**, 16997–17006.

17. T. Petrenko and F. Neese, Analysis and prediction of absorption band shapes, fluorescence band shapes, resonance Raman intensities, and excitation profiles using the time-dependent theory of electronic spectroscopy, *J. Chem. Phys.*, 2007, **127**, 164319.

18. T. Petrenko and F. Neese, Efficient and automatic calculation of optical band shapes and resonance Raman spectra for larger molecules within the independent mode displaced harmonic oscillator model, *J. Chem. Phys.*, 2012, **137**, 234107.

Chapter 6

Atomistic Simulation of Hierarchical Nanostructured Materials for Optical Chemical Sensing

6.1 Introduction

Materials for optical chemical sensing represent a gratifying subject for atomistic multiscale simulations, and we will try to use these materials in order to illustrate the strategy, main approaches, and main problems peculiar to atomistic multiscale simulations of materials for any other purpose, primarily for photonic applications. Here, we mainly follow Ref. [1].

Optical chemical sensors (OCSs) are designed to detect various chemical compounds (analytes) by a change in the optical properties of a sensing element (sensor) due to its interaction with an analyte. OCSs are widely used in various environmental, biomedical, and industrial applications (see, for example, Refs. [2, 3]).

The term "optical chemical sensors" was first used in the mid-1980s [3, 4]. Somewhat later, the use of dyes and nanofabrication for optical chemical sensing was reported [5–8]. The use of an array of OCS elements was also discussed in the literature starting from the end of the 1990s [9, 10] to the current decade [11]. In the past three decades, the number of studies in the field of OCSs has been growing rather quickly.

Multiscale Modeling in Nanophotonics: Materials and Simulation
Alexander Bagaturyants and Mikhail Vener

ISBN 978-981-4774-40-6 (Hardcover), 978-1-315-10976-3 (eBook)
www.panstanford.com

A wide range of sensing materials based on a dye immobilized on a nanoparticle can be created by combining different dyes, particles, and methods of immobilization. The immobilization of an indicator dye molecule on the particle surface significantly reduces the response and relaxation time of a sensing material [12].

The design and development of materials and devices for the detection and monitoring of volatile chemical compounds in various media that are comparable with olfactory systems of living organisms is an important problem. One of the solutions of this problem is to use an optical signal generated by a sensing material that changes its optical (absorption or emission) properties upon interaction with a target compound. The use of a matrix system composed of a set of different sensing elements allows one to analyze complex mixtures of volatile chemical compounds.

Materials with a broad spectrum of various prescribed characteristics of optical chemical sensing can be developed using a hierarchical theoretical design of nanostructured materials based on atomistic multiscale simulation, which involves the simulation of a supramolecular receptor center (RC) and its interactions with its local environment and with an analyte. Such an RC involves an absorbing or emitting indicator molecule (IM). Changes in spectral absorption or emission characteristics due to interaction of an RC with a gas-phase analyte molecule can be used for analyte detection. These RCs are either created directly or immobilized on an organic or inorganic nanoparticle. These modified nanoparticles are assembled into microstructures (nanoparticle assemblies) serving as elements of a matrix chemical sensor (chemochip) [13].

Therefore, predictive modeling of nanostructured materials is important for designing sensitive and selective OCSs. In particular, a multiscale approach is well applicable to the simulation of OCSs. In this approach, the calculations at each level are performed using methods and approximations peculiar to this level, and the results of simulating the structure and properties of the material obtained at a lower level are transferred as input data to the next upper one [13].

The use of a multiscale approach for the predictive atomistic simulation of materials and, especially, functional materials, was started in late 1990s (see, for example, Refs. [14, 15]), and these studies have given rise to numerous publications nowadays [16]. Actually, a multiscale atomistic approach was used in molecular simulation much earlier. Thus, already in 1990, classical mechanics was combined with semiempirical quantum chemistry to calculate the absorption line shapes of some organic chromophores, their dimers, and aggregates [17–19]. An integrated multiscale approach to atomistic simulation of film deposition processes in which a kinetic Monte Carlo (MC) method was combined with ab initio quantum-chemical (QC) calculations was described in a series of works [20–22].

In this chapter we will consider a multiscale atomistic approach to the predictive simulation of hierarchical nanostructured functional materials for OCSs.

6.2 Hierarchical Structure of Nanomaterials for Nanophotonics: From a Molecule to a Supramolecular Center, Nanoparticle, and Nanomaterial

An OCS is based on the effect of a change in the optical properties of an IM upon its interaction with an analyte. An IM is usually a dye molecule (or a different substance possessing specific spectral properties). Hence, the material of an OCS contains IMs, while a signal from the sensor is a change of its optical properties (positions and intensities of its spectral bands) upon its interaction with an analyte (optical response). Fluorescent optical sensors, in which the output signal is a change in the fluorescence properties of sensing materials [23], are among the most promising ones because of the high sensitivity of the fluorescence signal to various agents and, as a consequence, the possibility of detecting extremely low concentrations of analytes. While designing sensors, one should know the most important properties of the material, such as the energy of analyte interaction with the RC and changes in the spectral (absorption and emission) bands due to this interaction. Such properties of materials can be predicted using atomistic multiscale computer simulation.

At the lowest, molecular level, an IM is the main sensing element of such a material. The selection of the IM is determined by its ability to (selectively) interact with the analyte molecule and its ability to (selectively) change its optical properties (primarily, fluorescence intensity) upon this interaction.

At the next, supramolecular level, a supramolecular receptor center (SRC) should be considered. An SRC includes an IM and its nearest environment. This environment (the architecture of the SRC) is selected (designed) in such a way that the interaction between the analyte molecule and the SRC is optimized, the selectivity of this interaction is maximized, and the optical response of the designed SRC is enhanced as much as possible.

At the higher, nanosized level, a nanoparticle containing SRCs at its surface or in its bulk is created. The construction of a sensing material from nanoparticles provides an increased accessible surface for analytes, easy penetration of their molecules to RCs, and a possibility of creating a special structure (architecture) of the material with the best conditions for light scattering (or light absorption), for example, by creating a photonic lattice.

At the next hierarchical, micro level, a (2D or 3D) assembly of uniform nanoparticles ordered in a certain way is formed. This assembly contains IMs of only one certain type. This type is characterized by a certain value of the response to a certain analyte molecule. This assembly makes the basis for the designed material for OCSs.

The general task of multiscale simulation is the prediction and design of the material structure and response at each scale level. The results of simulation at a certain level are transferred to the next level, and the scheme of multiscale (multilevel) simulation is accomplished in this way. This scheme corresponds to a bottom-up design of the material. Actually, this scheme includes some feedback connections between different levels when the results of a higher level are used to refine the results of a lower level of modeling (top-down approach). This combined approach (bottom-up/top-down) is widely used for simulations and theoretical predictions of nanomaterials' structure and properties for a broad area of applications (optical chemical sensing, light-emitting and photovoltaic materials, materials for photonic crystals, optical memory media, etc.).

The main feature of this multiscale approach is that modeling starts from the lowest molecular level with the use of atomistic ab initio (first-principles) methods, based on fundamental laws of atomic and molecular interactions. At each next level, the results of the previous level are used as input parameters, while modeling is based on physical models and methods specially designed for the corresponding timescales and space scales.

6.3 Hierarchy of Atomistic Simulation Methods Corresponding to Scale Hierarchy

Simulation methods at a molecular or atomistic level are now well designed and implemented in a variety of commercial and noncommercial programs and program packages. Quantum-mechanical first-principles (ab initio) methods and programs differ substantially, depending on whether these are designed for the calculations of finite or infinite periodic systems. Conventionally, these methods might be classified, respectively, as molecular (cluster) or solid-state (periodic). It should be noted, though, that many molecular programs involve the capability of treating periodic systems, while solid-state programs can also treat molecular systems.

In the cluster approach, the surface of a nanoparticle is modeled by a finite atomic cluster, which is selected so that all the atoms that compose the RC (i.e., the IM and its nearest environment) are included. In most typical cases, the number of atoms in the RC does not exceed the limit acceptable for quantum calculations. In the periodic approach, the surface of a nanoparticle is modeled by a slab infinite and periodic in two dimensions with a certain finite thickness in the third direction (the so-called repeated-slab approximation [24]). To ensure the full 3D periodicity, the slab is periodically repeated in the third direction in such a way that two neighboring slabs are separated by a vacuum gap of a certain length (for more details, see, for example, Ref. [25]).

Simulation methods that can be used at the next dimensional (supramolecular) level are mostly based on the use of classical interaction potentials. These methods include molecular mechanics (MM), molecular dynamics (MD), and MC techniques with various sets of classical atomistic potentials or force fields, which are implemented in various program packages (see Chapters 3 and 4). At present, a number of integrated program packages exist, oriented to the calculation of materials properties based on atomistic approaches. These packages commonly include a variety of applied programs that can be used for modeling crystalline and molecular materials, biological objects, solutions, etc., combined by a well-developed user-friendly graphical interface. Information on the majority of available specialized and general-purpose programs, both commercial and noncommercial, can simply be found on the Internet.[a]

The efficient use of atomistic simulation can substantially cut down the time for the design of nanomaterials with desired properties and reduce their cost. As a rule, various requirements are imposed on the structure and properties of such materials, and atomistic simulation is necessary to obtain the best implementation of these requirements at the microscopic (molecular, supramolecular, and nanosized) levels. In many cases, supramolecular complexes (SMCs), that is, intermolecular complexes formed by two or more molecules or molecular fragments of the nanoparticle, are the main structure elements governing the most practically important properties of nanomaterials. Determining the possible configurations of SMCs for certain molecules and estimating the stability and required properties of the corresponding structures are among the most important tasks of atomistic simulation in the guided design of functional nanomaterials.

One of the main problems arising in determining the possible configurations of SMCs is connected with the necessity of an exhaustive search in the configurational space of the SMC. If the binding sites of the molecules are well defined or their structure is relatively simple, this search can be rather quickly performed manually using molecular visualization and editing tools. Generally, however, the number of possible configurations of an SMC drastically increases (e.g., in the case when a dye molecule is bound on a silica or polystyrene [PS] surface), and special techniques are necessary to search and select the most stable supramolecular configurations. To do this, various versions of global optimization algorithms are used [26] in which the global potential energy minimum is searched for an SMC calculated using classical force fields (e.g., [27]). The

[a]http://en.wikipedia.org/wiki/Molecular_modelling, http://en.wikipedia.org/wiki/List_of_software_for_nanostructures_modeling.

use of a genetic algorithm for this purposes combined with the use of a classical force field for the calculation of intermolecular interactions between the components of SMCs was also proposed and implemented in a computational program [28].

In some cases, the use of force fields to calculate the potential energy during global optimization can result in inaccuracies of both the geometry and relative energy of different configurations of an SMC. In particular, the geometry of conjugated aromatic rings typical in dyes is distorted when calculated with a force field. To refine the geometry of configurations, one should use QC methods. These methods make it possible to determine the geometry of an SMC and its important properties (absorption spectrum, luminescence spectrum, dipole moment, polarizability, etc.) from the first principles. Because QC calculations require considerable computational resources, they can be used for the local geometry optimization of an SMC; the structures obtained by global optimization should be used as starting geometries. For SMCs consisting of several hundreds of atoms, various versions of density functional theory (DFT) can be used for local optimization (see Chapter 3, Section 3.6). The interaction energy of the most important fragments of an SMC can be calculated using more accurate electron correlation methods, such as MP2 perturbation theory (see Chapter 3, Section 3.5.4; for a modern review, see Ref. [29]) and the coupled-cluster method (see Chapter 3, Section 3.5.5, and Ref. [30] for a recent review).

After determining the structure of the lowest energy configurations of an SMC, one should estimate their stability with regard to temperature. The free energy of formation ΔG_b, which can be calculated using MC or MD simulation, is a measure of the relative stability of intermolecular complexes. These methods make it possible to average the geometry (and some other properties) of an SMC over microstates with regard to temperature and to calculate ΔG_b from a MD or MC trajectory, using thermodynamic integration [31] or harmonic analysis [32]. Hence, it is evident that in modeling supramolecular structures, it is necessary to use a hierarchical approach, in which atomistic simulation techniques of different levels are employed to solve certain problems arising when different scales of the system are considered.

6.4 Atomistic Multiscale Simulation of Hierarchical Nanomaterials for Optical Chemical Sensors: Step by Step

6.4.1 Supramolecular Level: Calculations of Molecular Interactions between Gas-Phase Analyte Molecules and Simple Substrate Models

According to the general strategy for the hierarchical design of nanostructured optical sensing materials, an indicator dye molecule is adsorbed on the surface of a nanoparticle (which is used as a substrate). The nanoparticles are assembled into a sensing layer (sensor material), which can be used to construct a sensor array (for more details, see Refs. [13, 28, 33, 34]). An indicator dye molecule along with its local environment forms a functional SMC, whose properties depend on both the indicator dye itself and its environment. The interaction of an analyte molecule with the supramolecular center generates an optical response of the material. Therefore, a rational approach to the design of OCS materials should be based on a consideration of intermolecular interactions between analytes, indicators, and matrices [35].

Some simplified cluster models were used rather long ago in modeling interactions at polyene/graphite and polymer/polymer interfaces [36, 37] and more recently in modeling interactions of nitrile, aromatic, and olefinic polymers with silica [38].

In a general case, all types of interactions between these components must be taken into account. It may be believed that the possibility of detecting an analyte by an indicator depends on the mode and strength of interaction between the analyte molecule and the indicator. For the successful detection of low concentrations of an analyte, this interaction should be stronger than the interaction of the analyte with the substrate. Interactions of these types were investigated for some small molecules (possible analytes) interacting with PS and amorphous silica substrates and with an indicator dye molecule of the acridine series immobilized on these substrates using simplified molecular models of the substrate and the indicator dye [39–41]. Only the main active surface center (for the substrates) or the fragment (for the dye molecule) responsible for the interaction with analyte molecules was included in these models.

The ethylbenzene molecule was chosen in Refs. [39–41] as the simplest model of PS, which includes its main constituent moieties (phenyl ring and alkyl chain) and represents an elementary link of its polymer chain. The silanol molecule SiH_3OH was selected as the simplest model of a silanol group on the silica surface. The acridine molecule was taken as the main fragment of an acridine dye responsible for interaction with analyte molecules.

Formaldehyde, acetaldehyde, ammonia, methylamine, methanol, ethanol, acetone, benzene, acetonitrile, ethyl acetate, chloroform, and tetrahydrofuran were considered as the analyte molecules. QC calculations were made using DFT with an empirical dispersion correction (DFT-D) approach implemented in the ORCA program [42]. In this approach, an empirical correction for dispersion interaction was taken into account [43–46]. See the original paper [39] for more computational details.

The calculated lengths of shortest contacts in these complexes were used to analyze the corresponding coordination modes, while the interaction energies were used to compare the possibilities of detecting these small molecules, considered as analytes, using an acridine dye adsorbed on the surface of a PS or amorphous silica substrate. It was shown that most complexes of ethylbenzene exhibited a stacking structure in which the analyte molecule is located over the plane of the phenyl ring with the only exception for the ammonia complex. In the latter, the lateral coordination mode is preferable. All complexes of silanol and most complexes of acridine have a lateral structure formed through an Si–O–H···X H-bond (X is an analyte heteroatom) or an H(analyte)···N(acridine) H-bond.

The available experimental data on the heats of adsorption on PS and amorphous silica for some of the considered analytes indicated that the adsorbed molecules could interact simultaneously with more than one surface group. The calculated interaction energies indicate that the dyes of the acridine series adsorbed on a PS or silica substrate are not promising IMs for acetone and acetonitrile. For all other analyte molecules, PS could be considered a suitable substrate for an acridine dye indicator. Finally, silica might be considered a suitable substrate for the detection of methanol, ethanol, benzene, ethyl acetate, and chloroform using an acridine dye as a molecular sensor. The properties of the nanoparticles used as chemosensing materials can be improved by increasing the coverage of nanoparticles with indicator dye molecules.

6.4.2 Supramolecular Level: DFT Calculations of the 9-Diphenylaminoacridine Fluorescent Indicator and Its Interactions with Analyte Molecules

In a more complicated model, QC calculations were performed for a real functional dye molecule to study its interaction with analyte molecules and the effect of this interaction on the absorption and emission spectra of the dye [40, 41]. A dye of the 9-aminoacridine series, 9-diphenylaminoacridine (9-DPAA) [47], was selected for this study.

This dye was characterized by pronounced solvatochromism [48] and, therefore, was selected as an IM that can be used in molecular fluorescence sensors for the detection of polar organic molecules [35]. The first intense absorption and fluorescence bands of 9-DPAA were located in the visible spectral region, which simplified the detection. On the basis of the experimental data, it was assumed that the observed changes in the spectra were due to specific intermolecular interactions in complexes forming between 9-DPAA molecules and one or several analyte molecules. Theoretical calculations were performed for 1:1 complexes between molecules of typical volatile analytes and 9-DPAA [40, 41]. The resulting changes in the absorption and fluorescence spectra of the dye due to its interaction with some polar molecules (methanol, acetonitrile, acetone, tetrahydrofuran, ammonia, formaldehyde, and acetaldehyde) and nonpolar benzene were calculated.

A reasonable characterization of excited states was attained with the use of time-dependent density functional theory (TDDFT) (see Chapter 3, Section 3.6.4) with hybrid functionals, in particular PBE0, which was recommended for calculations of the electronic spectra of dyes [49], and with the 6-31G(d,p) basis set. For systems in the ground electronic states, geometry optimization was performed by the DFT method with the same functional and the same basis set. All calculations were performed by the Firefly program (Alex A. Granovsky, Firefly version 8, http://classic.chem.msu.su/gran/

firefly/index.html, Firefly QC package [50], which is partially based on the GAMESS-US source code) [51]. The GAMESS-US program package [51] was used for geometry optimizations in the excited state by the TDDFT method.

The results of QC simulations showed that all the analyte molecules under consideration (methanol, acetonitrile, acetone, tetrahydrofuran, benzene, ammonia, formaldehyde, and acetaldehyde) could form 9-DPAA complexes of two types: complexes with a lateral structure and complexes with a stacking structure. In the lateral complexes, the analyte molecule was located approximately in the plane of the acridine fragment of 9-DPAA on the side of the N atom and interacted with it through an X–H···N (X = O, C) H-bond. When the analyte molecule contained an electronegative atom (O or N), an additional contact arose between this atom and one of the positively charged H atoms of the acridine fragment. In stacking complexes, the analyte molecule was located over the plane of the acridine fragment. As well as lateral complexes, stacking complexes exhibited a short X–H…N contact with the acridine N atom; however, there was an additional short contact of the electronegative atom of the analyte molecule with the positively charged H atom of the benzene ring of the diphenylamino group.

Calculated vertical transition energies in the free 9-DPAA molecule with large oscillator strengths (at 463, 363, and 292 nm) corresponded well to experimental absorption bands with maxima at 450, 360, and 290 nm. The electronic transition between the ground and lowest excited singlet states was accompanied by a nearly complete transfer of an electron from the diphenylamino group to the acridine fragment, so the dipole moment of the molecule drastically increased from 1.9 D for the ground state to 15.4 D for the excited state. The corresponding experimental estimates for the methyl-substituted 9-DPAA analog 2,7-dimethyl-9-ditolylaminoacridine were 2.0 and 12 D [48]; that is, the TDDFT calculations overestimated charge transfer (CT) upon excitation (see Chapter 3, Section 3.7.2). The calculated absorption spectra of 9-DPAA complexes with the considered analytes insignificantly differ from the absorption spectrum of the free 9-DPAA molecule. All these results were in good agreement with experimental findings [48].

For the free 9-DPAA molecule, the calculated position of the fluorescence band at 588 nm was substantially larger than the experimental value of 490 nm for its analog 2,7-dimethyl-9(ditolylamino)acridine (9-DTAA) in a nonpolar hexane solution. The calculated Stokes shift was 125 nm, while the experimental value in solution was 40 nm [48]. Thus, calculations significantly underestimate the CT transition energy for the emission band, which was characteristic of TDDFT methods (see Chapter 3, Section 3.7.2, and Ref. [52]).

On the other hand, the calculated shifts of the positions of the fluorescence bands due to the formation of the stacking-type 9-DPAA complexes with the analytes correlated well with the experimental solvatochromic shifts of the fluorescence band in the corresponding solvent (R^2 = 0.90). To the contrary, the band shifts upon the formation of lateral complexes were small and did not correlate with the value of solvatochromic effects. Hence, solvatochromic effects in the case of 9-DPAA and 9-DTAA were controlled by specific solvation and the formation of the stacking structures of solvate complexes of the dye.

The structures of 9-DPAA complexes with the analytes were further refined by DFT-D calculations with the B97-D exchange correlation functional, in which a correction for dispersion interactions was directly included [53]. The inclusion of the dispersion correction significantly increased the relative stability of stacking complexes (by up to 6.2 kcal/mol). For almost all of the 9-DPAA-analyte complexes, the stacking structures were found to be more favorable than the lateral structures. The only exception was for the methanol complex, for which the lateral structure remained slightly more stable (by 1.6 kcal/mol) than the stacking one. The structures remained qualitatively unchanged in comparison with DFT calculations.

6.4.3 Multiscale Level: MD/DFT Slab Modeling of the Adsorption of Simple Organic and Inorganic Molecules on an Amorphous Silica Surface

Silica gel nanoparticles were widely used as a very promising substrate for hierarchical nanostructured materials for OCSs. It was known for its good adsorption capacity, porous structure, and large surface area (up to 1000 m^2/g); it also did not absorb light in the visible spectrum region (see, for example, recent reviews [54–56]).

As discussed before in this section, two different approaches are commonly used in modeling nanoparticle surfaces at a quantum-mechanical level: periodic (solid-state) and cluster (molecular). The periodic repeated-slab model was applied to studying the structure and electronic properties of an amorphous silica surface and its interaction with small organic and inorganic analyte molecules in Ref. [57]. In this work, the adsorption of simple molecules (water, ammonia, acetone, and ethanol) on the surface of a silica nanofilm was studied using the GGA DFT (see Chapter 3, Section 3.6.4) with the exchange correlation potential Perdew–Burke–Ernzerhof (PBE) [58] and with ultrasoft pseudopotentials [59]. The calculations were made using the quantum-mechanical program complex Quantum ESPRESSO [60].

The amorphous silica surface structure was simulated by classical MD using the interaction potential proposed in Ref. [61]. A $2 \times 2 \times 2$ supercell of α-quartz consisting of 72 atoms with the lattice period increased by 10% was heated up to 4000 K for 1000 ps. Then, it was held at 2500 K for 500 ps, and the obtained structure was further relaxed by periodic DFT calculations. The calculated densities of α-SiO_2 and amorphized silica were 2.6 g/cm^3 and 2.3 g/cm^3, respectively, in good agreement with the experimental densities.

The amorphized silicon oxide crystal cell was then used to construct a periodic slab. All the dangling bonds of the surface oxygen atoms were terminated with hydrogen atoms, thus forming surface silanol groups. Water, ammonia, acetone, and ethanol molecules were adsorbed on the optimized amorphous silica surface. Several positions of adsorption were found for each molecule. The calculated adsorption energies for the most stable adsorption positions varied from 14 to 18 kcal/mol for water, from 9 to 18 kcal/mol for ammonia, from 11 to 14 kcal/mol for ethanol, and from 7 to 8 kcal/mol for acetone. That is, water and ammonia were found to be more strongly adsorbed on the amorphous SiO_2 surface than ethanol and aceton.

6.4.4 Multiscale Level: MD/DFT Cluster Modeling of a 9-DPAA/Silica RC and Its Interaction with Small Analyte Molecules

The cluster approach with the DFT-D approximation (see Chapter 3, Section 3.6.4) was used in Ref. [62] for modeling the structure of an RC based on an 9-DPAA indicator dye adsorbed on the surface of amorphous silica particles (9-DPAA/silica) and adsorption of small analyte molecules (H_2O, NH_3, C_2H_5OH, and $(CH_3)_2CO$) on the surface of amorphous silica particles.

Different cluster models were used. The simplest ones were SiH_3OH (cluster **Si1**) and clusters **Si2–Si6** (containing from two to six Si atoms) artificially constructed from **Si1** by replacing hydrogen atoms with $Si(OH)_3$ groups, followed by full geometry optimization. The two larger clusters containing 10 and 20 Si atoms (**Si10**, **Si20**) were constructed using classical MD methods. MD simulations provide a most suitable tool for modeling amorphous silica clusters. Using this technique with corresponding force fields, one can amorphize the SiO_2 structure taking the structure of crystalline quartz as the initial approach, changing the unit cell dimensions, and performing simulation in the *NVT* ensemble. In the cited work, **Si10** and **Si20** clusters were constructed in a similar manner as in Ref. [57].

With increasing silica cluster size from **Si1** to **Si6**, the average energy of the intermolecular interaction between the dye molecule and the silica cluster increased until the number of accessible silanol groups reached five. Hence, it was sufficient to have approximately five or more silanol groups in contact with the dye for the adequate description of the dye/silica interaction energy.

Clusters **Si10** and **Si20** were composed of SiO_4 tetrahedra with Si–O bond lengths varying in a range of 1.60–1.66 Å; alternating Si and O atoms in the inner silica skeleton formed six-membered rings with Si–O–Si angles varying in a relatively wide range (130°–160°). Along with conventional six-membered rings, the **Si20** cluster also contained one strained four-membered Si–O–Si–O cycle with much smaller Si–O–Si angles (~108° and ~100°). The surfaces of both **Si10** and **Si20** particles were composed of Si–OH groups arranged randomly and linked via Si–O–Si bridges. The obtained surface structure was in a fairly good agreement with various theoretical models and some experimental data (see, for example, Refs. [63–65] and references therein)

The adsorption properties of amorphous silica species are determined by silanol groups, which exhibit acid properties and can interact with Lewis base molecules bearing lone electron pairs. Siloxane bridges, which involve alternating positively (Si) and negatively (O) charged atoms, can participate in electrostatic interactions with polar species. An amorphous silica particle, as a whole, possesses sufficient polarizability and, therefore, can participate in dispersion interactions with aromatic species.

To take into account all these interactions properly, silica clusters should be relatively large (in our case, of the order of 10.5 Å, the size of the dye molecule). Thus, the **Si10** cluster was a little smaller than, while the **Si20** cluster was approximately as large as the 9-DPAA molecule.

It was found that the analyte molecules were attached to the surface by relatively short H-bonds (formed by the lone electron pair of their donor N or O atoms with surface silanol groups) and by one or two longer H-bonds between analyte H atoms and O atoms of accessible silanol groups. The analyte adsorption energies depend on the local environment of the adsorbate molecule on the cluster surface. At the best computational level, the calculated adsorption energies of analytes on SiO_2 were found to be approximately 8–9 kcal/mol for H_2O and C_2H_5OH, 11–13.5 kcal/mol for NH_3, and 12–15 kcal/mol for $(CH_3)_2CO$.

The dye formed a relatively strong H-bond with one of the silanol groups by its acridine nitrogen atom. The calculated energy of 9-DPAA adsorption on SiO_2 was found to be from 2 to 3 times higher (22–29 kcal/mol) than the adsorption energies of the analytes. The presence of the dye molecule on the surface had no significant effect on the energies of analyte adsorption on SiO_2. The simultaneously adsorbed analyte molecule also had no significant effect on the adsorption of the dye. The QC calculations of the 9-DPAA/silica RC will be considered in more detail in the next section.

Thus, the results of this combined MD/QC study provided the geometry of an indicator center and the energies of molecular adsorption on the surface of small silica nanoparticles with a reasonable accuracy at a moderate computational cost. It was shown that the adsorption energy for the large 9-DPAA molecule depends on the cluster size and the number of accessible Si–OH groups; calculations with small cluster models give a somewhat underestimated binding energy for the 9-DPAA molecule. For small analyte molecules, the energies of adsorption on silica nanoparticles substantially (within several kcal/mol) depend on the adsorption position, and different adsorption sites with extended cluster models should be considered. The dye is much more strongly adsorbed than the small analyte molecules on the surface of amorphous silica nanoparticles. Hence, analytes could not displace the dye from the particle surface, so sensors based on adsorbed dyes of the 9-DPAA type will possess sufficient chemical stability to analyte vapors.

6.4.5 Multiscale Level: MD/DFT Cluster Modeling of the Effect of Analyte Molecules on the Absorption and Fluorescence Spectra of a 9-DPAA/Silica RC

The investigation of the 9-DPAA/silica/ RC was further extended in Ref. [66] to include more analyte molecules and to study the effect of their interaction with the RC on the RC fluorescence and absorption spectra. Both polar and nonpolar compounds were considered as analytes: acetone, ammonia, methanol, ethanol, water, benzene, naphthalene, toluene, and dinitrotoluene. The theoretical modeling of the indicator/matrix system included QC calculations of the interaction energies (DFT-D) and the energies and oscillator strengths of electronic transitions (TDDFT) in the dye molecule adsorbed on a model silica and in the analyte/RC systems. Models used in the cited work were similar to those used in Ref. [62].

The model of amorphous silica was constructed by classical MD simulations. A hexagonal α-quartz unit cell consisting of nine atoms was built, its dimensions were increased to reproduce the amorphous silica density (the densities of α-quartz and amorphous silica are 2.6 and 2.26 g/cm^3, respectively), and a 7 × 7 × 7 supercell was constructed (about 3000 atoms). The potentials were taken from Feuston and Garofalini (1988, 1990a, 1990b). This supercell was heated for 7 ps at 1 fs steps to the temperature 6000 K and then kept at 4000, 2000, 1000, and 300 K for 7, 7, 17, and 17 ps, respectively. This procedure gave an amorphous bulk SiO_2 structure.

The simplest model of silica was SiH_3OH (cluster **Si1**). The 9-DPAA/**Si1** complex was constructed by forming an H-bond between the nitrogen atom of the acridine fragment of the dye molecule and the silanol group of SiH_3OH (the nitrogen atom of the diphenylamine fragment is sterically inaccessible). Extending the cluster by a sequential substitution of H atoms with H–$Si(OH)_3$ groups showed that the cluster should contain about 10 silanol groups, and four of them should directly interact with the dye, to describe the 9-DPAA–cluster interaction correctly.

For QC calculations, a cluster of the least necessary size (**Si10**, 10 atoms Si, 57 atoms altogether, including the added –OH groups) was cut from the obtained bulk structure; the broken bonds of O atoms (so-called dangling bonds) were saturated with H atoms; the positions of thus obtained hydroxyl groups were optimized by DFT-D calculations (PBE0+D, 6-31G**, GAMESS-US). In optimizing the geometry of the cluster, all atoms except for the atoms of newly formed hydroxyl groups were kept frozen at their bulk positions (cluster atoms not interacting with the dye were also kept frozen in the further calculations of complexes with the **Si10** cluster).

For the cluster obtained, the best position was found for the attachment of the dye by maximizing both the number of –OH groups interacting with the dye and the projection of the dye onto the cluster (maximum contact area). The latter condition was used in order that the adsorption of the dye on the cluster better described its adsorption on the silica surface. The geometry of the system was finally optimized using the DFT-D approximation (B97-D, 6-31G**); the interaction energy was refined using the extended Dunning cc-pVTZ triple-zeta basis set with the geometries optimized with the double-zeta basis set. The energy of dye adsorption on the silica cluster was as high as 21.6 kcal/mol.

The most probable configurations of analyte complexes with the model RC were found using the INDAM program [28] implemented in the NanoModel software complex (NanoModel 2.3, http://www.nanomodel.ru). The INDAM program allows a search to be performed for the global energy minimum for a system of two interacting molecules without valence bonds (VBs) between them. The obtained structures were finally optimized by DFT-D calculations (PBE0+D/6-31G**, GAMESS-US).

To estimate the effect of the extension of the silica cluster on the electronic absorption spectrum of the RC and its complexes with small analyte molecules (acetone, ammonia, ethanol), large models based on the **Si10** cluster were also constructed. The structures of complexes were optimized within the DFT-D approximation (PBE0+D/6-31G**).

To determine the fluorescence maximum of the adsorbed dye, the geometry of the **Si1**/DPAA complex in its first S1 excited state was optimized by TDDFT calculations (PBE0/6-31G**, GAMESS-US). To make the calculations of the fluorescence spectra of 9-DPAA feasible, the **Si10** cluster was reduced by removing all –OH and Si–OH groups that do not directly contact with the dye; the broken bonds were terminated with hydrogen atoms. The further computational details may be found in the original paper [66].

The calculated position of the first absorption bands indicated that, for almost all systems (except for naphthalene), the interaction of the RC with the analytes is accompanied by small (of about 10–30 nm) red shifts of the absorption band. For naphthalene the shift of the absorption band is negligibly small. The calculated fluorescence spectra were in a reasonable agreement with the experimental fluorescence spectra of 9-DTAA adsorbed on spherical silica particles of the diameter 10 μm. From the calculated positions of low-lying triplet levels (T0 and T1) for the **Si1**/9-DPAA RC and **Si1**/9-DPAA/analyte (analyte = acetone, ammonia, water, and ethanol) complexes, it was found that the T0 level is considerably lower than the S1 level. In the **Si1**/9-DPAA RC, the T1 level is below S1 and rather close to it. This arrangement is favorable to fluorescence quenching. However, the interaction with the analytes changed the relative positions of S1 and T1 levels in the 9-DPAA/SiH_3OH complexes so that the T1 level becomes markedly higher than S1. Thus, interactions with analytes in the 9-DPAA/silica RC should prevent fluorescence quenching and lead to rise in fluorescence intensity, which was actually observed in the experiments [34].

Thus, it was shown by theoretical simulations that the 9-DPAA/silica RC can be used to detect vaporous acetone, ammonia, methanol, ethanol, benzene, and toluene by changes in fluorescence and absorption spectra.

6.4.6 Multiscale Level: Modeling the Structure and Spectra of an RC Based on the Nile Red Dye Adsorbed on the Surface of Polystyrene

Models of the structures of complexes of the Nile red (NR) dye molecule on the surface of various types of PS were constructed using MD simulations in Ref. [67]. The molecule of the dye consists of a rigid aromatic fragment and a relatively mobile diethylamino group, which can rotate around the C–N bond, adopting various positions.

The aim of this work was to model the structures of the NR/PS RC and to analyze the accessibility of the dye to the analyte in the composition of such a center. The molecule of the NR dye was suggested as a fluorescent IM in Ref. [68] and used in Ref. [34] and quite recently in Ref. [69] as a sensing element. In the latter work, PS was directly used as a matrix.

An all-atom approach was used in Ref. [67] to fully retain the chemical structure for both the matrix and the dye molecule during the modeling of their complex. All the MD simulations were made using the GROMACS package [70, 71] and the optimized potentials for liquid simulations–all atom (OPLS-aa) force field [72] for the PS and NR inter- and intramolecular interaction potentials. The initial geometry and the charges of the NR dye molecule were obtained from TDDFT QC calculations (B3LYP/6-31G*). The surface area for NR adsorption was found to be small and approximately equal to 1.4 nm^2. It was assumed that the surface structure of the polymer with such a small area had enough time to relax during the modeling time, while it was not necessary to attain equilibrium in the entire volume of the polymer. To construct a PS film, a system of 10 linear PS chains containing 50 monomer links each was used (500 monomer links altogether). This system was first thermalized at 500 K to remove the initial anisotropy. The PS density was deliberately lower than the experimental value, which facilitated the relaxation and mixing of the chains. After thermalization, the system was subjected to long-time shrinkage in the normal temperature and pressure (NPT) ensemble to reach the experimental density 1060 kg/m^3.

When a cubic PS cell was obtained, the periodic boundary conditions along the z axis were removed, and the size of the cell in this direction was restricted by the two repulsive walls. These walls were moved slowly together, so the system in the *xy* plane existed in an NTP ensemble, while its size was variable and depended on the pressure. The cubic volume of the polymer became flattened and adopted the form of a planar structure. At the end, the walls moved away so that one of them restricted the polymer from below while the other was 10 nm above it, leaving a vacuum of sufficient size above the polymer layer. This method of film modeling was described in detail in Ref. [73].

Harmonic confining potentials with a force constant of 1000 kJ/mol·nm^2 acting along the *z* axis were applied to the heavy atoms of the main chain of PS in the central part of the film. As a result, the PS bulk as a whole remained in place during the subsequent calculations, while the motion of separate sections of the chains was not restricted. After the final thermalization, the monomers at the PS surface on the vacuum side relaxed, and the film adopted the final form.

A globule was constructed in the cited work using an isolated polymer chain contained 500 monomer links as the initial structure. The system was heated to 500 K, thermalized for a few nanoseconds, and cooled down to 298 K. As a result, the chains rolled up into a globule of an imperfect spherical form. The NR molecule was initially located close to the PS surface, and then the NR/PS system was relaxed at T = 298 K to a constant average NR/PS interaction energy.

The structure of the PS surface layer in both models (film and globule) differed significantly from the bulk structure. The density of the polymer along the normal to the surface had the form of a plateau with a value close to the experimental value, which gradually decreased to zero in the surface area. The transverse length of the transitional surface layer in the film model was about 5 Å, which corresponded approximately to the size of the phenyl group. The adsorbed NR molecule penetrated partly into the space between PS links. This position corresponded to a local energy minimum. Thus, a quasi-equilibrium configuration was obtained for the dye in the environment of PS links. The amine fragment remained above the surface and had enough space for the rotation of the amine group. The last circumstance is important because the amine group in the excited state can rotate out of the plane of the aromatic system [74].

An analysis of the solvent-accessible surface (SAS) [75, 76] indicated that in the film model, an NR molecule adsorbed on the PS surface and buried in the space between PS links is accessible for a typical analyte molecule.

The PS density inside the globule was appreciably lower than inside the film, while the thickness of the transitional layer was larger. The NR molecule penetrated into the globule surface by the aromatic fragment deeper than in the film model and formed a strong complex with PS. However, as in the case of the film, the NR amine group remained in the free space accessible for interaction with analyte molecules and allowed the rotation of the ethyl groups around the C–N bond.

Thus, the MD simulations of the PS surface using both the periodic film model and the globule model showed that the dye molecule should partly penetrate into the PS surface and form a stable complex with surrounding segments of polymer chains. The morphology of the polymer surface had an effect on the sensor characteristics of the material, and this effect can be investigated by MD. The resulting sensitivity of the sensor material depended on the distribution of the investigated substrate/chromophore surface complexes over the degree of chromophore accessibility to a certain analyte. In principle, the obtained sensitivity may differ for analytes of different size and form, which provided the basis for predicting the selectivity of a potential sensor material.

On the basis of the results obtained in Ref. [67], the structure and absorption spectra of the NR dye adsorbed on the surface of a PS nanoparticle was studied by DFT in Ref. [77]. A PS nanoparticle with an adsorbed NR molecule was simulated by a large cluster consisting of PS chains using MD in the OPLS-aa force field [72] in the *NVT* ensemble at 298 K. Different surface models were tried: a surface of a single-chain coil (up to 1000 monomeric units) and of a periodic box. Next, the surface was truncated so that only the nearest-neighbor phenyl rings surrounding the chromophore were treated either explicitly by DFT or as effective fragment potentials (EFPs) proposed by Jensen et al. [78], a recent review of EFPs and their applications was done in Ref. [79]. The structure of the dye obtained in Ref. [67] from MD calculation was reoptimized by DFT in the frozen environment, and its absorption and emission spectra were calculated by TDDFT in this environment and compared with those in vacuo and in toluene (simulated through the polarizable continuum model [PCM]). It was shown that the transition energies calculated using an explicit environment and using EFPs are almost the same, but the use of EFPs strongly reduces the time of calculations.

6.4.7 Prospects and Outlook

The general considerations and various examples presented above demonstrated that the multiscale atomistic modeling of nanomaterials for OCSs allowed one to predict the structure of an RC, its interactions with analyte molecules, and the spectral response of the RC to these interactions. These results provided a basis for a directional search for new, most promising materials for optical chemical gas-phase sensing, possessing the best sensitivity and selectivity.

An important new step toward a more comprehensive description of the optical response of a nanomaterial was made quite recently in Ref. [80]. In this work, a simple and convenient approach was proposed to evaluating band shapes in the electronic spectra of complex molecular systems from first principles based on the classical multiphonon model [81]. In this model, each electronic transition is broadened into a structureless band of approximately Gaussian shape due to the linear electron-vibration coupling. The required parameters for these model calculations can be calculated by QC methods, such as DFT and TDDFT, as described in detail in Ref. [80], so the bandwidths and shapes of interest can easily be calculated from the results of QC calculations.

This rather promising approach was quite recently applied to modeling the band shapes in the electronic absorption spectra of a silica/DPAA RC and its complexes with analytes in Ref. [82]. A set of both polar and nonpolar analytes included acetone, ammonia, methanol, ethanol, water, benzene, naphthalene, toluene, and dinitrotoluene. The results of calculations of the bands were in good agreement with the experimental data for DPAA in solutions, which led the authors to the conclusion that the developed approach can be used for the estimation of the shapes of spectral bands in the absorption spectra of organic dyes adsorbed on silica particles and their complexes with analytes.

6.5 Simulation of Amorphous Substrates: Polymers, Small Molecule Layers, and Silica

Ab initio calculations at the MP2 level of theory with a 6-31 G basis set that included polarization functions on the atoms involved in the H-bond, were performed for the hydrogen-bonded phenol-NH_3 complex. The equilibrium O···N distance, R_e, was equal to 2.837 A. The potential profile along the proton coordinate at a fixed O···N distance had only one minimum. Its shape was verified by calculation of the OH stretch frequency. The shape of the $V(r, R_e)$ function for systems with an O–H···N fragment in the gas phase was reproduced using the Lippincott–Schroeder potential only if some its parameters were calibrated to fit high-level ab initio data [83].

Temperature dependence of the proton spin-lattice relaxation time (T_1) in a powdered benzoic acid dimer and in its deuterated analog was calculated. The model assumed that two protons (deuterons) synchronously move in the double-minimum potential of the dimer. The 2D potential energy surface constructed previously adequately described the static properties of the hydrogen-bonded complex. The important characteristics of this potential were a very strong mode coupling and a very high proton potential barrier (>25 kcal/mol), whereas the experimental activation energy for proton transfer (PT) was known to be of the order of 1 kcal/mol only. This apparent discrepancy was removed by the suggestion that PT was driven by transitions between vibrational levels of the OHO fragment under the action of random forces of the surrounding. The excitation of low-frequency intermolecular vibrations strongly assisted this transfer mechanism. Using four fitting parameters to take into account medium repolarization, the calculated T_1 temperature dependence was found to be in good agreement with experiments in the natural and deuterated benzoic acid dimer. The agreement was best at high temperature, where the apparent activation energy for PT was found to be 2.3 kcal/mol [84].

The adiabatic separation of the vibrational variables of a hydrogen-bonded AHA fragment with a symmetric double-well potential was considered using two approaches. In the first approach, the A···A stretch vibration was found to be slow in comparison to the proton vibration. It was shown that this approach was applicable to the energy level calculations of an AHA fragment with a potential barrier of an arbitrary height. In the alternative approach, the proton was considered to be a slow subsystem. This approach appeared to be of limited applicability. It was shown that, for an AHA fragment with a symmetric double-well potential, the Born–Oppenheimer criterion contained the frequency corresponding to the proton 0–2 transition in both approaches [85].

A local continuum solvation theory, exactly treating electrostatic matching conditions on the boundary of a cavity occupied by a solute particle, was extended to cover time-dependent solvation phenomena. The corresponding integral equation was solved with a complex-valued frequency-dependent dielectric function $\varepsilon(\omega)$, resulting in a complex-valued ω-dependent reaction field. The inverse Fourier transform then produced the real-valued solvation energy, presented in the form of a time correlation function (TCF). This technique was applied to describe the solvation TCF for a benzophenone anion in Debye (acetonitrile) and two-mode Debye (dimethylformamide) solvents. For the Debye solvent, the TCF is described by two exponential components and for the two-mode Debye solvent, by three. The overall dynamics in each case was longer than that given by the simple continuum model. A steady-state kinetic regime was also considered and the corresponding rate constant for adiabatic electron transfer (ET) reactions. Here the boundary effect introduced within a frequency-dependent theory generates only a small effect in comparison with calculations made within the static continuum model [86].

TCFs describing the solvent relaxation around a molecule of coumarin-153 and a benzophenone anion in acetonitrile were calculated using dynamical continuum theories of solvation with an experimental dielectric function $\varepsilon(\omega)$, including the resonance absorption region of the solvent. In addition to the local model with a single molecular-shaped solute cavity of the solute studied previously, a new dynamic local model with a double molecular-shaped cavity and a dynamic nonlocal theory with a spherical cavity were presented, both of which introduced elements of the solvent structure. It was shown that both local models, one- and two-cavity, exhibited experimentally unobserved oscillations in the shorter time region $t < 1$ ps, although the experimental asymptote for $t > 1$ ps was obtained for coumarin. The dynamics of the two-cavity model did not differ from those of the one-cavity model. The nonlocal dynamic was able to suppress these oscillations, but the long-time asymptote differed markedly from that of the local theories. The nature of this asymptote was studied analytically [87].

A linearized MD technique was developed to treat systematically free energies of simple charged and dipolar solutes in water. The solvent electrostatic response field in the solute region was modeled by averaging instantaneous fields found in an MD computation for solvent configurations confined within a cavity that conforms to the real shape of the solute particle. At this stage, all electrostatic interactions were explicitly treated inside the cavity. The solvent in the external region (outside the cavity) was modeled in terms of standard continuum theory. For nonspherical cavities, this approach was more accurate than the field computation employed at the preceding MD stage, where spherically truncated Coulomb potentials are modified by the reaction field corrections. Two different linearization schemes were considered based on a computation of either the average response field or its fluctuations. Only the first algorithm proved to be successful. For a series of single-charged monatomic cations and anions, it provided free energies, which deviated by few percentage points from those found in full MD computations. The results were stable relative to a separation of the whole space occupied by the solvent into explicit solvent region (inside the cavity) and the continuum region (outside the cavity). The two-site dipolar dumbbell system was also studied in the range of intersite separation D within 2 Å < D < 10 Å. At the stage of the field computation, three different types of its solvation shell were considered: spherical and bispherical cavities and periodic solvent environment monitored in terms of the Ewald method. Free solvation energies were the same (within 1 kcal/mol) for all three models. A smooth dependence of the mean-field potential was observed as a function of separation D, but its asymptotic value differed by 4 kcal/mol from the free energy computed for the isolated ion pair [88].

A combined molecular-continuum approach developed in Ref. [89] was applied to calculations of equilibrium solvation energies for a large number of polyatomic ions. The structure and charge distribution of a given ion were computed using the restricted Hartree–Fock (HF) level with the 6-31G** basis set. The standard Lennard–Jones (LJ) parameters, which were not specially calibrated to fit the solvation energies, were used in MD simulations. Water (the SPC model) was considered as a solvent. The computations showed that the new scheme worked satisfactorily for nitrogen cations with a standard parametrization. This scheme could be further improved for oxygen ions by tuning solute-solvent LJ parameters. The calculated relative change of the energies in families of similar (ammonium-type or oxonium-type) cations was in agreement with experimental trends [90].

The hybrid molecular-continuum model for polar solvation, which combined the dielectric continuum approximation for treating fast electronic (inertialess) polarization effects and MD simulation for the slow (inertial) polarization component, including orientational and translational solvent modes, was considered. The inertial polarization was generated by average charge distributions of solvent particles, composed of permanent and induced (electronic) components. MD simulations were performed in a manner consistent with the choice of solvent and solute charges such that all electrostatic interactions were scaled by the factor $1/\varepsilon(\infty)$, where $\varepsilon(\infty)$ was the optical dielectric permittivity. This approach gave an ensemble of equilibrium solvent configurations adjusted to the electric field created by a charged or strongly polar solute. The electrostatic solvent response field was found as the solution of the Poisson equation, including both solute and explicit solvent charges, with accurate account of electrostatic boundary conditions at the surfaces separating spatial regions with different dielectric permittivities. Both equilibrium and nonequilibrium solvation effects could be studied by means of this model, and their inertial and inertialess contributions were naturally separated. A methodology for the computation of charge-transfer reorganization energies was developed and applied to a model two-site dipolar system in the SPC water solvent. Three types of CT reactions are considered. The standard linear response approach yielded high accuracy for each particular reaction but proved to be significantly in error when reorganization energies of different reactions were compared. This result has a purely molecular origin and is absent within a conventional continuum solvent model [89].

The closed-shell interactions in oligo- and polyalanines were studied by the quantum theory of atoms in molecules (QTAIM) using electron densities derived from the B3LYP/6-31+G** ground-state electronic wave functions. The intrachain H-bonds between the adjacent N–H and C=O groups in the antiparallel β-sheet conformation of polyalanine were not found, while these interactions do exist in the parallel conformation, being even stronger than the interchain N–H···O bonds [91].

The structure, harmonic frequencies, and infrared (IR) intensities of the fundamental transitions in the potassium hydrogen maleate (KHM) crystal have been computed using the BLYP/6-31G** approximation with periodic boundary conditions. The inelastic neutron scattering (INS) intensities of the fundamental transitions were evaluated and compared with the experimental INS spectra of hydrogenated sample and isotopic analogues of the KHM crystal: the agreement is reasonable good. Spectroscopic manifestations of the proton dynamics in the strong intramolecular H-bond were clarified by comparison of the computed frequencies of the IR- and INS-active fundamental transitions. Due to several reasons, the number of the bands and their relative intensity differed between the INS and IR spectra of the KHM crystal. Reduced masses for the asymmetric O···H···O stretching vibrations of the hydrogen maleate ion appeared to be larger than 1 amu because of a strong coupling between the bridging proton motion and various intramolecular vibrations [92].

Proton forms of zeolite chabazite (H-SSZ-13) loaded with 1 to 4 water molecules per acid site were examined by DFT with periodic boundary conditions. Equilibrium structures were determined by localizing minima on the potential energy surface, and harmonic vibrational frequencies were calculated. Average structures, proton dynamics and anharmonic spectra at finite temperature (350 K) were determined by MD. The protonation state was found to be dependent on the number of water molecules per acid site (loading), following the trend of increasing proton affinity with increasing cluster size. Single water molecules were not protonated, and the protonated water dimer was the most stable equilibrium structure with the PBE functional, but not with BLYP. MD showed that even with PBE, the protonated water dimer was not stable at finite temperature. The protonated water trimer can be formed as a short-lived species, but the protonated water tetramer was the smallest stable protonated cluster. For the same global loading (2:1), a heterogeneous distribution of adsorbed water molecules over the cells is more stable than a homogeneous one (1:1/3:1 versus

2:1/2:1 for a double cell), that is, nonprotonated and protonated water clusters may exist simultaneously in polyhydrated H-SSZ13. Adsorption energies (0 K) per water molecule decrease from 71 to 51 kJ/mol for $n = 1$ to $n = 4$ [93].

The alanine (Ala)-based cluster models of C5, C7, and C10 H-bonds were studied at the DFT/B3LYP level. Car–Parrinello molecular dynamics (CPMD)/Becke–Lee–Yang–Parr (BLYP) simulations of an infinite polyalanine α-helix (C13 H-bond) and the two-stranded β-sheets were performed. Combined use of frequency shifts and electron density features enable us to detect and describe quantitatively the noncovalent interactions (H-bonds) defining the intrinsic properties of Ala-based secondary structures. The energies of the primary N–H···O H-bonds are decreasing in the following way: C13 > C5 >= C7 > C10. The energies of the secondary N–H···O, N–H···N, and H···H interactions are comparable to those of the primary H-bonds (similar to 4.5 kcal/mol). Side-chain–backbone C–H···O interaction was found to be the weakest noncovalent interaction in the considered species. Its energy is similar to 0.5 kcal/mol in the infinite polyalanine α-helix. Quantum-topological electron density analysis was found to be a powerful tool for the detection of secondary noncovalent interactions (C=O···H–C and H···H) and bifurcated H-bonds, while the frequency shift study was found to be useful for the identification and characterization of primary or secondary H-bonds of the N–H···O type [94].

The geometric parameters and relative stability of γ- and β-loop conformers of oligomers based on alanine were determined by the Kohn–Sham (KS) method (the B3LYP/6-31+G** approximation). The 3D architecture of the β-folded structure and protein α-helix was reproduced using KS calculations with periodic boundary conditions. The Bader quantum-topological molecular structure theory was used to reveal and quantitatively characterize noncovalent interatomic interactions in the secondary structures of model peptides under consideration. Additional noncovalent interactions stabilizing the structures under consideration, which were not noted previously, were revealed. Additional weak bonding interaction between C=O groups and H–C(beta) groups was revealed for the protein α-helix; this interaction is usually ignored in amino acid–folding simulations with the use of classical force fields [95].

The backbone–side-chain interactions in the peptide secondary structures were studied by DFT methods with/without periodic boundary conditions. Alanine-based two-stranded β-sheet structure infinite models and cluster models of C5 structures modified by the glutamic acid residue were considered. Several low-energy structures were localized in the BLYP/plane wave and the BLYP/6-311++G** approximations. Noncovalent interactions and H-bonds were quantitatively described on the basis of a combined use of a quantum-topological analysis of the electron density and frequency shifts. It was found that the strongest backbone–side-chain interaction (about 37 kJ/mol) was due to the intrachain H-bond formed by the C=O backbone group and by the COOH side-chain group. The OH···O distance was found to be 1.727 Å and the frequency shift of the OH stretching vibration was found to be 370 cm^{-1}. The polar side-chain interaction was studied in the infinite model of the alanine-based two-stranded β-sheet structure modified by the glutamic acid/lysine residues. A moderate interchain H-bond (about 40 kJ/mol) was formed by the glutamic acid COOH group and the lysine NH_2 group. The OH···N distance equaled to 1.707 Å and the frequency shift of the OH stretching vibration was 770 cm^{-1} [96].

The O–H/H...O bond order indices were evaluated at the B3LYP level using the quantum theory of atoms in molecules and crystals (QTAIMC). The parameters of bond critical points, such as electron density, principal electron density curvatures, and kinetic and potential energy densities, were calculated for 32 molecular systems containing 40 covalent O–H bonds and 69 hydrogen H···O bonds of different strength. Different models for the bond order index were tested using the Cioslowski-Mixon approach as a standard. A uniform description of bond order indices for covalent and H-bonds in the O–H···O fragment ranging from 0.97 to 2.91 Å was achieved using the electron density and electronic kinetic energy density features. A model describing the H-bond indices of the X–H···O fragment (X = O, N, and C) in terms of electron density and potential energy density was also suggested [97].

Hydrogen bonding in ionic liquids based on the 1-(2′-hydroxylethyl)-3-methylimidazolium cation ([C(2)OHmim](+)) and various anions ([A](−)) of differing H-bond acceptor strength, viz. hexafluorophosphate [PF6](−), tetrafluoroborate [BF4](−), bis-(trifluoromethanesulfonimide) [Tf2N](−), trifluoromethylsulfonate

[OTf](−), and trifluoroacetate [TFA](−), was studied by a range of spectroscopic and computational techniques and, in the case of [C(2)OHmim][PF6], by single-crystal X-ray diffraction. The first quantitative estimates of the energy (E_{HB}) and the enthalpy ($-\Delta H_{HB}$) of H-bonds in bulk ILs were obtained from a theoretical analysis of the solid-state electron density map of crystalline [C(2)OHmim] [PF6] and an analysis of the IR spectra in crystal and liquid samples. The E_{HB} for OH···[PF6](−) H-bonds amounts to similar to 3.4–3.8 kcal mol^{-1}, whereas weaker H-bonds (2.8–3.1 kcal mol^{-1}) are formed between the aromatic C2H group of the imidazolium ring and the [PF6](−) anion. The enthalpy of the OH···[A](−) H-bonds follows the order [PF6] (2.4 kcal mol^{-1}) < [BF4] (3.3 kcal mol^{-1}) < [Tf2N] (3.4 kcal mol^{-1}) <[OTf] (4.7 kcal mol^{-1}) < [TEA] (6.2 kcal mol^{-1}). The formation of aggregates of self-associated [C(2)OHmim](+) cations is present in liquid [C(2)OHmim][PF6], [C(2)OHmim] [BF4], and [C(2)OHmim] [Tf2N], with the energy of the OH···OH H-bonds amounting to similar to 6 kcal mol^{-1}. Multiple secondary interactions in the bulk ILs influence their structure, vibrational spectra, and H-bond strength. In particular, these interactions can blue-shift the stretching frequencies of the CH groups of the imidazolium ring in spite of red-shifting CH···[A](−) H-bonds. They also weaken the H-bonding in the IL relative to the isolated ion pairs, with these anticooperative effects amounting to ~50% of the E_{HB} value [98].

A continuum model for solvation effects in binary solvent mixtures was formulated in terms of DFT. The presence of two variables, namely the dimensionless solvent composition y and the dimensionless total solvent density z, is an essential feature of binary systems. Their coupling, hidden in the structure of the local dielectric permittivity function, was postulated at a phenomenological level. Local equilibrium conditions were derived by a variation of the free-energy functional expressed in terms of the composition and density variables. They appear as a pair of coupled equations determining y and z as spatial distributions. The simplest spherically symmetric case of the Born-type ion immersed in a benzene/dimethylsulfoxide (DMSO) solvent mixture. The profiles of $y(R)$ and $z(R)$ along the radius R, which measures the distance from the ion center, were found in MD simulations. It was shown that, for a given solute ion, $z(R)$ does not depend significantly on the composition variable y.

A simplified solution is then obtained by inserting $z(R)$ found in the MD simulation for pure DMSO as a single equation that defines $y(R)$. The composition dependence of the main solvation effects was investigated. The local density augmentation appeared as a peak of $z(R)$ at the ion boundary. It is responsible for fine solvation effects missing when ordinary solvation theories in which $z = 1$ are applied. These phenomena, studied for negative ions, reproduce consistently the simulation results. For positive ions, the simulation showed that $z > 1$ ($z = 5 \div 6$) (at the maximum of the $z(R)$ peak), which means that an extremely dense solvation shell is formed. In such a situation, the continuum description is not valid within a consistent parametrization [99].

A new qualitative approach for treating theoretically the solvation of single-atomic ionic solutes in binary mixtures of polar and nonpolar aprotic solvents was considered. It is based on the implicit continuum electrostatic model of the solvent mixture involving distance-dependent dielectric permittivity $\varepsilon(R)$ (where R is the distance from the ion) and local concentrations $C(1)(R)$ and $C(2)(R)$ of the solvent ingredients. For a given R, the condition for local thermodynamic equilibrium provides the transcendental equation for explicitly establishing the permittivity and concentration profiles. Computations performed with real Cl(−) and model Cl(+) ions as solutes in benzene/DMSO mixtures are compared with the MD simulations of the same systems. A significant discrepancy of molecular and continuum results is revealed for the concentration profiles in the close vicinity of the ion boundary, although the general trends are similar. The continuum methodology cannot account for the formation of rigid solvent structures around ions, which is most significant for the case of Cl(+). Such defect, however, proves to become of less importance in calculations of the solvation free energy, which are quite satisfactory for Cl(−) ion. Free-energy calculations for Cl(+) are less successful in the range of low DMSO concentration [100].

Potentials of mean force (PMFs) were calculated for two model ion pairs in two nonaqueous solvents. Standard nonpolarizable molecular dynamics (NPMD) simulation and approximate polarizable molecular dynamics (PMD) simulation were implemented and compared as tools for monitoring PMF profiles. For a polar solvent (dimethylsulfoxide, DMSO) the PMF generated in terms of the NPMD reproduced fairly well the refined PMD-PMF profile. For the nonpolar solvent (benzene), the conventional NPMD computation proved to be deficient. The validity of the correction found in terms of the approximate PMD approach was verified by its comparison with the result of the explicit PMD computation in benzene. The shapes of the PMF profiles in DMSO and in benzene were quite different. In DMSO, owing to dielectric screening, the PMF presented a flat plot with a shallow minimum positioned in the vicinity of the van der Waals contact of the ion pair. For the benzene case, the observed minimum proves to be unexpectedly deep, which manifests the formation of a tightly bound contact ion pair. This remarkable effect arises because of the strong electrostatic interaction, which is incompletely screened by a nonpolar medium. The PMFs for the binary benzene/DMSO mixtures display intermediate behavior depending on the DMSO content [101].

Models of the structures of complexes of the NR dye molecule on the surface of various types of PS were constructed by MD. The surface of a PS film and the surfaces of clumps of isolated PS chains were examined as surfaces. The film and the clumps were obtained as a result of MD trajectories. The atomistic version of MD was used. The surface of the film was comparatively uniform, and the clumps were irregular ellipsoids with a varied local surface form. The variety of the surface forms makes it possible to obtain complexes of PS with local environments having different structures. A method of constructing the solvent-accessible surface (SAS) was proposed as method of evaluating the potential sensor characteristics of the material. In the PS clumps, the NR molecule is almost completely submerged in the upper layer, while in the film the aromatic fragment of the dye is partly accessible to the analyte [67].

The double-decoupling version of the thermodynamic integration procedure was applied to perform the molecular dynamical modeling of binding free energies of β-cyclodextrine (CD) with a number of organic ligands. Simulations for water solutions show a satisfactory agreement (within 12 kcal/mol) with the experimentally measured equilibrium binding constants. The values were also reported for the gas-phase complexation of the same ligands, although no experimental data are available for such systems. These gas-phase computations revealed a large stabilization effect for the CD complexes of ionic ligands. Only in this special case, the attempt of qualitatively rationalizing the obtained simulation data proved to be fairly successful. The problems specific for simulations for ionic ligands in water solution were discussed [102].

A molecular dynamic approach was applied to the simulation of dynamics of pore formation and growth in a phospholipid bilayer in the presence of an external electric field. Processing the simulation results permitted the recovery of the kinetic coefficients used in the Einstein–Smoluchowski equation describing the dynamics of pore evolution. Two different models of the bilayer membrane were considered: membrane consisting of phosphatidylcholine (POPC) and phosphatidylethanolamine (POPE) lipids. The simulations permitted us to find nonempirical values of the pore energy parameters, which were compared with empirical values. It was found that the parameters are sensitive to the membrane type [103].

The results of a qualitative analysis of the effect of a counterion and polar environment on the structure and spectroscopic signatures of a hydrated hydroxyl anion based on DFT/M06-2X calculations in the polarizable cavity approximation were reported. These results provide a tool to discriminate between the spectral signatures of the neutral OH·(nH(2)O) B+, and anionic, OH·(nH(2)O), solvation clusters in aqueous solutions of alkali metal (B) hydroxides. These results demonstrate also that the precisions of the available experimental data and of the double harmonic approximation are insufficient to attribute the experimental spectral signatures to any specific cluster with the certain number of involved water molecules at least for $3 <= n <= 6$ [104].

Salt bridges and ionic interactions play an important role in protein stability, protein-protein interactions, and protein folding. Here, we provide the classical MD simulations of the structure and IR signatures of the arginine (Arg)-glutamate (Glu) salt bridge. The Arg-Glu model is based on the infinite polyalanine antiparallel two-stranded β-sheet structure. NPT simulations show that it preferably exists as a salt bridge (a contact ion pair). Bidentate (the end-on and side-on structures) and monodentate (the backside structure) configurations were localized [105]. These structures are stabilized by short N^+–H . . . O bonds. Their relative stability depends on the force field used in MD simulations. The side-on structure is the most stable in terms of the OPLS-aa force field. If AMBER ff99SB-

ILDN is used, the backside structure is the most stable. Compared with experimental data, simulations using the OPLS-aa force field describe the stability of the salt bridge structures quite realistically. It decreases in the following order: side-on > end-on > backside. The most stable side-on structure lives several nanoseconds. The less stable backside structure exists a few tenth of a nanosecond. Several shortliving species (solvent shared, completely separately solvated ionic groups ion pairs, etc.) are also localized. Their lifetime is a few tens of picoseconds or less. Conformational flexibility of amino acids forming the salt bridge is investigated. The spectral signature of the Arg-Glu salt bridge is the IR-intensive band around 2200 cm^{-1}. It is caused by the asymmetric stretching vibrations of the N^{+}–H···O fragment. These results suggest that infrared spectroscopy in the 2000–2800 frequency region may be a rapid and quantitative method for the study of salt bridges in peptides and ionic interactions between proteins. This region is usually not considered in spectroscopic studies of peptides and proteins [106].

Viscoelastic properties of a molecular liquid consisting of 4,4′-*N*,*N*′-dicarbazolylbiphenyl (CBP) molecules near the glass transition temperature were investigated by MD simulations. The relaxation dynamics was analyzed by considering each molecule as a point-like oriented particle. The dependence of the calculated properties on the coarse-grain parameter used in the calculation of orientation correlation was analyzed. The divergence of α-relaxation times is described by the Vogel–Fulcher–Tammann law and the mode coupling theory. The basic concepts of the glass transition theory were applied to a real amorphous organic semiconductor [107].

Many key features of the protein adsorption on the silica surfaces still remain unraveled. One of the open questions the interaction of nonpolar side chains with siloxane cavities. Nonequilibrium MD simulations were used for the detailed investigation of the binding of several hydrophobic and amphiphilic protein side chains with silica surface. These interactions were found to be a possible driving force for protein adsorption. The free-energy gain was larger for the disordered surface of amorphous silica gel as compared to α-quartz; but the effect depended on the type of amino acid. The dependence was analyzed from the structural point of view. For every amino acid, an enthalpy-entropy compensation behavior was observed. These results confirm a hypothesis of an essential role of hydrophobic interactions in protein unfolding and irreversible adsorption on the silica surface [108].

Benzene and other nonpolar organic solvents can accumulate in the lipid bilayer of cellular membranes. Their effect on the membrane structure and fluidity determines their toxic properties and antibiotic action of the organic solvents on the bacteria. MD simulations of the interaction of benzene with the dimyristoylphosphatidylcholine (DMPC) bilayer was performed. An increase in the membrane surface area and fluidity was clearly detected. Changes in the acyl chain ordering, tilt angle, and overall bilayer thickness were, however, much less evident. The dependence of all computed quantities on the benzene content showed two regimes separated by the solubility limit of benzene in water. When the amount of benzene exceeded this point, a layer of almost pure benzene started to grow between the membrane leaflets. This process corresponds to the nucleation of a new phase and provides a molecular mechanism for the mechanical rupture of the bilayer under the action of nonpolar compounds [109].

6.6 Guest Molecules in Amorphous Hosts

A method for introducing an effective coordinate for the solvation shell in a hydrogen-bonded cluster with the A–H···B reaction complex was presented. Because of the formation of additional hydrogen bonds between the A–H···B complex and the molecules of the solvation shell, PT can occur. A model 3D potential energy surface (PES) that included explicitly the O–H and O···N stretching modes was extracted from an ab initio surface. The PES was calculated for a cluster containing a phenol-ammonia complex surrounded by an ammonia shell $(NH_3)_4$. The two lowest proton adiabatic terms were computed. Two possible approaches for treating PT dynamics in the two-level approximation were discussed [110].

Intrinsic solubility of the isoniazid-carboxylic acid cocrystals is explored using thermodynamic cycle of transferring from crystal to gas and then to solution. The enthalpy/Gibbs free energy of phase transitions is computed by solid-state DFT (crystal) and DFT methods with/without accounting for a polar solvent (aqueous solution/gas). The dissolution process of the isoniazid-dicarboxylic acid 2:1 cocrystals is found to be enthalpy determined. The linear correlation between the theoretical sublimation enthalpies and experimental Gibbs energy of solution is established. According to it, the increase of stability causes the decrease of the considered cocrystals solubility. We conclude that a soluble and relatively stable cocrystal of isoniazid should be characterized by the sublimation enthalpy varying from 185 to 200 kJ/mol. Bader analysis of theoretical periodic electron density enables us to quantify the non covalent interactions and clarify their role in formation of the isoniazid cocrystals. The sublimation enthalpy of the cocrystals is defined by the structure of the coformer molecule. Its value is less 150 kJ/mol in the case of aromatic acids and larger than 180 kJ/mol for the aliphatic dicarboxylic acids. The sublimation enthalpy increases with the increase of the dicarboxylic acid hydrocarbon chain [111].

Cocrystal screening of 4-hydroxybenzamide with a number of salicylates (salicylic acid, SA; 4- aminosalicylic acid, PASA; acetylsalicylic acid, ASA; and salicylsalicylic acid, SSA) was conducted to confirm the formation of two cocrystals, [SA +4-OHBZA] (1:1) and [PASA+4-OHBZA] (1:1). Their structures were determined using single-crystal X-ray diffraction, and the hydrogen bond network topology was studied. Thermodynamic characteristics of salicylic acid cocrystal sublimation were obtained experimentally. It was proved that PASA cocrystallization with 4-OHBZA makes the drug more stable and prevents the irreversible process of decarboxylation of PASA, resulting in the formation of toxic 3-aminophenol. The pattern of noncovalent interactions in the cocrystals was described quantitatively using solid-state DFT followed by Bader analysis of the periodic electron density. It has been found that the total energy of secondary interactions between synthon atoms and the side hydroxyl group of the acid molecule in [SA+4-OHBZA] (1:1) and [PASA+4-OHBZA] (1:1) cocrystals is comparable to the energy of the primary acid amide heterosynthon. The theoretical value of the sublimation enthalpy of [SA+4-OHBZA], 231 kJ/mol, agrees fairly well with the experimental one, 272 kJ/mol [112].

The lattice energy E_{latt} of the two-component crystals (three cocrystals, a salt, and a hydrate) was evaluated using two schemes. The first one was based on the total energy of the crystal and its components computed using the solid-state DFT method with the plane-wave basis set. The second approach explored intermolecular energies estimated using the bond critical point parameters obtained from the Bader analysis of crystalline electron density or the pairwise potentials. The E_{latt} values of two-component crystals were found to be lower or equal to the sum of the absolute sublimation enthalpies of the pure components. The computed energies of the supramolecular synthons vary from similar to 80 to similar to 30 kJ/mol and decrease in the following order: acidamide > acidpyridine > hydroxylacid > amideamide > hydroxylpyridine. The contributions from different types of noncovalent interactions to the E_{latt} value were analyzed. It was found that at least 50% of the lattice energy comes from the heterosynthon and a few relatively strong H-bonds between the heterodimer and the adjacent molecules [113].

The structure, IR harmonic frequencies and intensities of normal vibrations of 20 molecular crystals with the X–Cl···Cl–X contacts of different types, where X = C, Cl, and F and the Cl···Cl distance varies from similar to 3.0 to similar to 4.0 Å, were computed using the solid-state DFT method. The obtained crystalline wave functions have been further used to define and describe quantitatively the Cl···Cl interactions via the electron density features at the Cl···Cl bond critical points. It was found that the electron density at the bond critical point was almost independent of the particular type of the contact or the hybridization of the ipso carbon atom. The energy of Cl···Cl interactions, E_{int}, was evaluated from the linking E_{int} and the local electronic kinetic energy density at the Cl···Cl bond critical points. E_{int} varied from 2 to 12 kJ/mol [114].

Noncovalent interactions of different types and strengths in the crystal of picolinic acid N-oxide were studied experimentally and theoretically based on the charge density distribution. The energies of the intramolecular O–H···O bond, intermolecular C–H···O bonds, and π-stacking interactions (E_{int}) were evaluated from empirical correlations between E_{int} and geometrical and electron density bond critical point parameters. The complete set of noncovalent interactions including the strong intramolecular O–H···O (E_{int} > 90 kJ/mol) and weak CH··· O (E_{int} < 10 kJ/mol) hydrogen bonds, and π-stacking interactions (E_{int} < 4 kJ/mol) were quantitatively described. The results from the experimental charge density analysis were compared with periodic quantum calculations using

DFT with the Grimme dispersion correction. It was found that the Grimme dispersion correction did not provide a good simultaneous description of both weak and strong noncovalent interactions in the studied crystal. It was shown that the obtained energies of noncovalent interactions lead to a reasonable value of the lattice energy [115].

The hydrogen bond (H-bond) energies were evaluated for 18 molecular crystals with 28 moderate and strong O–H···O bonds using the approaches based on the electron density properties, derived from the B3LYP/6-311G** calculations with periodic boundary conditions. DFT computations with periodic boundary conditions and precise X-ray diffraction experiments [116].

The structure, IR spectrum, and H-bond network in the serine-H_2O and serine-H_2O_2 crystals were studied using DFT computations with periodic boundary conditions. Two different basis sets were used: the all-electron GTO basis set and the plane wave basis set. The quantum-topological analysis of the crystalline electron density was used to describe quantitatively the H-bond network [117].

The crystal structure of $NH_4^+OOH^-$ was determined from single-crystal x-ray data obtained at 150 K. The crystal belonged to the space group P2(1)/c and had four molecules in a unit cell. The structure consisted of discrete NH_4^+ and OOH^- ions. The OOH^- ions were linked by short hydrogen bonds (2.533 Å) to form parallel infinite chains. The ammonium ions form links between these chains (the N···O distances vary from 2.714 to 2.855 Å) giving a 3D network. The harmonic IR spectrum and H-bond energies were computed at the PBE/6-31G** level with periodic boundary conditions. A detailed analysis of the dynamics of shared (bridging) protons was obtained from the CPMD simulations at different temperatures. The PBE functional with plane-wave basis set (110 Ry) is used. At 10 K, the shared proton is located near the oxygen atom; only a few proton jumps along the chain were detected at 70 K, while numerous proton jumps exist in the trajectory at 270 K. The local-minimum structure of the space group Cc was localized. It appeared as a result of PT along a chain. This process was endothermic (similar to 2 kJ/mol) and was described as P2(1)/c <-> 2Cc. The computed IR spectrum at 10 K is close to the harmonic one, the numerous bands appear at 70 K while at 270 K it shows a very broad absorption band that covers frequencies from about 1000 to 3000 cm^{-1}. The advantages of the $NH_4^+OOH^-$ crystal as a promising model for the experimental and DFT-based MD simulation studies of PT along the chain are discussed [118].

The structure, harmonic frequencies, and infrared intensities of the fundamental transitions of potassium hydrogen/deuterium maleate crystals have been computed by DFT with periodic boundary conditions. Different functionals with all-electron Gaussian-type orbital (GTO) basis set have been used. It was found that BLYP/GTO approximation provides the best results for the structural parameters of the KHM crystal. Within this approximation, the hydrogen-bonded potential in the KHM crystal is extremely shallow. Delocalization of the bridging protons complicates a strict definition of the space group of the crystal; the space groups Pbcm and Pbc2(1) represent an equivalent choice. The periodic BLYP/GTO study provides detailed information on the nature of the internal vibrations of the hydrogen maleate anion, which are located in the range 300–1800 cm^{-1}. Assignment of the vibrational bands in this frequency region has been performed. The two most intensive bands in the infrared spectrum (similar to 500 and similar to 1450 cm^{-1}) are caused by the pure asymmetric O···H···O stretching vibrations and the stretching motion of the bridging proton heavily mixed with the C-C stretching and CH bending vibrations, respectively. A crystalline environment is found to play an important role in the low-frequency region and is negligible above 1000 cm^{-1}. The H/D substitution slightly changes the vibration frequencies, involving the stretching motion of the bridging proton because of the strong coupling between this motion and the various internal vibrations of the hydrogen maleate anion [119].

The characteristics of the critical electron density points of the O–H···A fragment (A = O or N) in molecular crystals with short H-bonds were analyzed in terms of the Bader quantumtopological theory of molecular structure. The wave functions (B3LYP/6-31G** approximation) of the ground state of 26 3D periodic crystals with experimentally determined structures were used. Intermediate-type interactions separating the limiting cases of covalent and closed shell-type interactions were found to be characterized by the following geometric parameters: 2.45 Å <= D(O···O) <= 2.6 Å, 1.35 Å <= d(H···O) <= 1.65 Å, and 1.0 Å <= d(O–H) <= 1.10 Å. Such interactions are observed in molecular crystals with the O–H···A fragment and high bridge proton mobility. Differences between H···O and H···N interactions were quantitatively characterized by the dependence of rho (b) on d(H···A), where A = O or N. The dependence parameter values were shown to be determined by the nature of the A atom that forms the H-bond. The influence of the crystalline environment on systems with strong H-bonds manifested itself by changes in the position of the bridge proton. The d(O–H)/d(H···O) ratio was, however, the same for gas-phase complexes and molecular crystals with weakly bent O–H···O fragments (angle O–H···O > 160 degrees) [120].

Conformational analysis of cumene hydroperoxide PhCMe2OOH (1) was carried out using the density functional methods B3LYP/6-31G(d,p) and B3LYP/6-311+G(3df,2p). Ignoring rotation of methyl groups, molecule 1 has seven conformers differing in the orientation of the –CMe2OOH fragment relative to the benzene ring and in mutual position of atoms in this fragment. The molecular structures, relative energies, and statistical distribution of the conformers were determined, and intramolecular rotational barriers were estimated. The enthalpies of formation of all conformers of molecule 1 were calculated using two approximations with inclusion of zero-point vibrational energy and temperature correction. Calculations using the isodesmic reaction (IDR) scheme made it possible to reduce the systematic error of the determination of the enthalpy of reactions. The total enthalpy of formation of compound 1 calculated with inclusion of statistical distribution of rotamers equals –19.7 +/– 3.6 kcal/mol. The combination of the B3LYP/6-31G(d, p) approximation and the IDR scheme gives fairly accurate results (relative error is +/– 0.4 kcal/mol) as compared to those obtained with the extended basis set 6-311+G(3df, 2p) [121].

The relationship between the d(H···A) distance (A = O, N) and the topological properties at the H···A bond critical point of 37 strong (short) hydrogen bonds occurring in 26 molecular crystals were analyzed using the QTAIM. Ground-state wave functions of the 3D periodical structures representing the accurate experimental geometries calculated at the B3LYP/6-31G** level of approximation were used to obtain the QTAIM electron density characteristics. The use of an electron-correlated method allowed quantitatively correct values of electron density at the H···A bond critical point (rho(b)) to be obtained. However, quite significant differences can appear for small absolute values of the Laplacian (<0.5 au). The difference between the H···O and H···N interactions was described using the rho(b) versus d(H···A) dependence. It was demonstrated that the values of parameters in this dependence were defined by the nature of the heavy atom forming the H···A bond. An intermediate (or transit) region separating the shared and closed-shell interactions was observed for the H-bonded crystals in which the bridging proton could move from one heavy atom to another. The crystalline environment changed the location of the bridging proton in strong H-bonded systems; however, the d(O–H)/d(H···O) ratio was approximately the same for both the gas-phase complexes and molecular crystals with a linear or near-linear O–H···O bond [122].

Second-order Moller–Plesset perturbation theory and DFT were employed to localize several stationary points on the potential energy surface of the cyclic methanol tetramer. Two cyclic isomers were identified: one of S_4 symmetry, with methyl groups in up-down-up-down configuration, and a second one of C-i symmetry, with the methyl groups in an up-up-down-down configuration. The latter minimum is 360 cm^{-1} above the S_4 minimum, with a barrier of 475 cm^{-1} separating them. These isomers gave rise to four asymmetric OH modes around 3300 cm^{-1}. A model of the concerted PT, $S_4 \rightarrow$ D-2d $\rightarrow S_4$ (D-2d is the transition structure), yielded an estimate of 0.7 cm^{-1} for the tunneling splitting of the totally symmetric OH stretch vibrational fundamental. Raman spectra would show evidence of this fundamental and help to identify admixtures of the C-i isomer [123].

6.7 Absorption and Emission Spectra, Line Shapes

The effect of a high-affinity peptide (HAP) homologous to a fragment of the nicotinic acetylcholine receptor (nAChR) on the absorption and fluorescence spectra of thiazole orange and thioflavin T complexes with cucurbit[8]uril in aqueous solution was analyzed. In the presence of HAP, a change in the absorption spectra of the dye complexes and a drop in the fluorescence intensity occured; for thiazole orange, the fluorescence intensity was restored to the initial level in the presence of α-bungarotoxin capable of high-affinity binding to nAChR. The proposed method made it possible to detect the presence of α-bungarotoxin [124].

The structure and absorption spectra of the Nile red (NR) dye adsorbed on the surface of a PS nanoparticle was studied by DFT/TDDFT based on the results of previous MD simulations. From NR/PS surface structures obtained from MD simulations, relatively small clusters were constructed composed of the adsorbed NR molecule and the nearest PS fragments. These clusters were treated either explicitly by DFT/TDDFT or using the EFP model. The absorption and emission spectra of the NR/PS system calculated using the EFP model are very close to those calculated with the explicit environment, so the EFP model can be recommended for use in multiple large-scale calculations. The calculated positions of the absorption and fluorescence bands were in good agreement with experiment [77].

The values of steady-state solvatochromic Stokes shifts (SS) in absorption/emission electronic spectra of organic chromophores were studied theoretically in the framework of the Hush–Marcus model. Charge distributions for chromophore solutes in their S_0 and S_1 states were found by means of conventional QC methods combined with the PCM approach for treating solvation effects. The solvent reorganization energies, which were expected to correlate with the solvent-induced part of 1/2 SS, were found in an MD simulation which invoked a novel method for separation of the inertial piece of the electrostatic response [125a]. Computations, performed in two solvents (acetonitrile and benzene), consider three organic dyes: coumarin 153 as a benchmark system and two other chromophores, for which experimental spectra were also reported. The results were found to be in reasonable agreement with the experiment. A consistent treatment of the nonlinear effect in the solvent response, promoted by the polarizability of solutes and contributing to the solvent reorganization energies [125b], improved the results of computations [126].

The spectrum of guanidinium acetate in aqueous solution was recorded by attenuated total reflectance infrared (ATR-IR) spectroscopy. Assignments of the bands were done using the PCM. Three IR-intensive bands at 1670, 1550, and 1410 cm^{-1} were associated with stretching and bending vibrations of the groups forming a ring of six heavy atoms of the bidentate configuration of guanidinium acetate. The relatively weak broad band near 2200 cm^{-1} was tentatively assigned to the stretching vibration of the N–H···O fragment of the hydrogen-bonded ion pairs [127].

Salt bridges and ionic interactions play an important role in protein stability, protein-protein interactions, and protein folding. Classical MD simulations of the structure and IR signatures of the arginine (Arg)-glutamate (Glu) salt bridge were performed. The Arg-Glu model was based on the infinite polyalanine antiparallel two-stranded β-sheet structure. The 1 μs NPT simulations showed that it preferably existed as a salt bridge (a contact ion pair). Bidentate (the end-on and side-on structures) and monodentate (the backside structure) configurations were localized [105]. These structures were stabilized by short N+–H···O– bonds. Their relative stability depended on the force field used in the MD simulations. The side-on structure was found to be the most stable in terms of the OPLS-aa force field. If the AMBER ff99SB-ILDN field was used, the backside structure was the most stable. Compared with experimental data, simulations using the OPLS-aa force field described the stability of the salt bridge structures quite realistically. It decreased in the following order: side-on > end-on > backside. The most stable side-on structure lived several nanoseconds. The less stable backside structure existed a few tenth of a nanosecond. Several short-living species (solvent shared, completely separately solvated ionic groups ion pairs, etc.) were also localized. Their lifetime was a few tens of picoseconds or less. Conformational flexibility of amino acids forming the salt bridge was investigated. The spectral signature of the Arg-Glu salt bridge was the IR-intensive band around 2200 cm^{-1}. It was caused by the asymmetric stretching vibrations of the N+–H···O fragment. This result indicated that infrared spectroscopy in the 2000–2800 frequency region might be a rapid and quantitative method for the study of salt bridges in peptides and ionic interactions between proteins. This region was usually not considered in spectroscopic studies of peptides and proteins [106].

The IR spectra of water-ethanol (H_2O-EtOH) solutions of HCl were measured over a wide range of acid concentrations at fixed H_2O-EtOH ratios (1:1, 1:2, and 1:40). In these systems, different proton disolvates with (quasi)symmetrical H-bonds were formed. Their structures and vibrational features were revealed by the DFT method coupled with the PCM of solvation. In dilute acidic solutions, the Zundel-type H_5O_2+ ion is mainly formed. In concentrated HCl solutions, the (H_2O···H···O(H)Et)(+) and (Et(H)O···H···O(H)Et)(+) ions with the quasi-symmetrical O···H···O unit having O···O separation <2.45 Å appear. The first ion characterized by the IR-intensive band around 1800 cm^{-1} was mainly formed in the 1:1 water-ethanol systems. The second ion existed in the 1:2 and 1:40 water-ethanol systems. Its spectroscopic signatures were groups of IR-intensive bands around 800 and 1050 cm^{-1}. In highly concentrated HCl solutions with the 1:40 water-ethanol ratio, a neutral Et(H)O···H+...Cl^- complex existed [128].

The structures, relative stability, and IR and Raman spectra of the most-stable forms of [H_9O_4](+) and [$H_{13}O_6$](+) ions in acetonitrile were computed using the B3LYP functional combined with the PCM approximation. These forms were hydrated [H_3O](+) and [H_5O_2](+) cores. Of interest were two main environmental effects on the spectroscopic features of protonated water hydrates: (i) polarization of the solvent by the hydrate dipole moment and (ii) formation of H-bonds with bulky counterions (ClO_4^- and BF_4^-). The effect of the polarization on the structure of the [H_3O](+) core strongly depended on the symmetry of the hydration shell. A distortion of a hydrated [H_3O](+) easily changed its structure to the [H_7O_3](+) one, which caused a change in the nature of the most IR-intensive bands. Thus, the specificity of this core can easily be lost, which prevented identification of the corresponding species. By contrast, the [H_5O_2](+) core is more stable against distortion. It was characterized by the short O···O distance (<2.45 Å), IR-intensive band near 1720 cm^{-1}, and Raman-intensive line around 500 cm^{-1}. The [H_5O_2](+) core remained identifiable even when a protonated hydrate was involved in specific interactions with a bulky counterion. Geometrical criteria for identification of the [H_3O](+), [H_5O_2](+) and [H_7O_3](+) cores were discussed [129].

Frequencies and IR intensities of the OH stretching vibrations in proton-ordered ices VIII, II, IX, and XI were computed using B3LYP/6-311G** with periodic boundary conditions. In accordance with the experiment, the average downshift of the OH or OD frequencies had a trend in changing from small to large shifts in the order: VIII < II < IX < XI. The H-bond enthalpy values evaluated from the IR intensities of the OH stretching vibrations were consistent with this trend. PBE/6-311G** and BLYP/6-311G** overestimated the OD frequency downshift. The bond paths in the periodic electron density, which are associated with the O···O interactions in ices VIII, II, and IX, were found [130].

The structure and spectroscopic properties of the 1:1 complexes of substituted pyridines with benzoic acid and phenol derivatives in aprotic solvents were studied using the B3LYP functional combined with the PCM approximation. Two extreme structures were investigated: the state without (HB) and with PT. In the presence of an external electric field, the O···N distance was contracted, and the PT state did appear. The PT state of both the pyridine-benzoic and the pyridine-phenol complexes displayed the only IR-active band in the 2800–1800 frequency region, which was located around 2000 cm^{-1}. However, the nature of the band was different for these two complexes. In the pyridine-benzoic acid complex, it was virtually a pure stretching vibration of the HN^+ group, while in the pyridine-phenol complex, it was the mixed vibration of the bridging proton. A specific feature of the PT state in the pyridine-phenol complex was an IR-intensive band near 600 cm^{-1}, associated with the asymmetric stretching vibrations of the $O \cdots HN^+$ fragment. Its intensity is reciprocally proportional to the O···N distance. The appearance of this band provides an efficient criterion to differentiate between the HB and PT states of the 1:1 complexes of phenols with pyridines in aprotic solvents [131].

The structure, IR spectra, and INS spectra of $H_5O_2^+$ in crystalline $H_5O_2{+}ClO_4^-$ were simulated using CPMD with the BLYP functional. The potential of the $O \cdots H^+ \cdots O$ fragment was found to be very shallow. The Pnma structure, assumed in the X-ray study to be the most suitable choice, was found to be a saddle point on the potential energy surface, while the P2(1)2(1)2(1) minimum structure was found to be only 20 cm^{-1} lower in energy. The computed INS and IR spectra made it possible to achieve a complete assignment of the observed spectra. The broad band between 1000 and 1400 cm^{-1} was found to be due to the asymmetric stretch and one of the bending vibrations of the $O \cdots H^+ \cdots O$ fragment, while the band between 1600 and 1800 cm^{-1} was due to the bending vibration of the water molecules and the second bending of the $O \cdots H^+ \cdots O$ fragment. Comparison with the vibrational spectra of isolated $H_5O_2^+$, obtained using Born–Oppenheimer MD simulation, revealed environmental effects on vibrational proton dynamics in strongly H-bonded species. The most pronounced changes were found for the O···H$^+$···O bending modes because the two bending coordinates become distinctly different for the structure that the $H_5O_2^+$ ion assumed in the crystal [132].

The vibrational spectrum of the formic acid dimer in the OH stretch region was calculated by solving a 3D vibrational Schrödinger equation and employing an electronic potential and dipole moment calculated at the MP2 level of theory. Two Raman-active bands of the symmetric OH stretch fundamental were located around 2600 cm^{-1}: the gap between them (similar to 300 cm^{-1}) was caused by concerted proton tunneling. IR-active bands around 3100 cm^{-1} were assigned to the antisymmetric OH stretch fundamental, its first overtone, and two complex transitions. The calculated HID isotope effect on the intensity of the antisymmetric OH fundamental is 1.92. The gas-phase IR spectrum of the O···H···O fragment of H_5O_2+ and its deuterated analogue are calculated using ab initio classical MD based on a MP2 potential energy surface. The assignment of the bands is made in terms of the quantum 4D model calculations of anharmonic frequencies and intensities. Comparing low- and high-temperature kinetic simulations the importance of anharmonicities of the potential energy surface for understanding the vibrational band structure is highlighted. It is shown that any reasonable simulation of IR spectra of systems with very strong hydrogen bonds has to account for the dipole moment function beyond the linear approximation [133].

Anharmonic frequencies of the O···H···O fragment of H_5O_2+ were evaluated using the adiabatic separation of vibrational variables. The relaxed 3D potential energy surface was calculated by the MP2 method using the T(O)DZP basis set. The symmetric and asymmetric stretching frequencies obtained, 588 and 1226 cm^{-1}, agreed with previous 2D results and showed a modest shift from the harmonic values only. The O···H···O bending frequency was ~1025 cm^{-1}, significantly lower than the harmonic result of 1524 cm^{-1}. Comparison was made with results of MD simulations using ab initio forces [134].

6.8 Charge and Excitation Transfer

The local hopping step of the electron transfer (ET) reaction was investigated for a real organic material composed of molecules M

(*N,N′*-di(1-naphthyl)-*N,N′*-diphenyl-(1,1′-biphenyl)-4,4′-diamine). This material is implemented in light-emitting photoelectronic devices. The conductivity effect was simulated and calculated at a molecular level. The ET mechanism alternative to that suggested by the usually employed Marcus-like polaron model was suggested. The ion-molecular binary complex MM (for hole transfer) was considered as a reaction center. The reaction dynamics was carried through the low-frequency intermolecular vibration coordinate connecting its fragments (the promotion mode). Its coupling to the acoustic phonon bath served for a dissipation of the reaction energy misfit. The high-frequency intramolecular vibrations (the reorganization modes) modulated the reaction kinetics via Franck–Condon factors induced by their polarization. The ET rate constants were evaluated in terms of the computational algorithm described earlier [135]. Standard QC and MD techniques were used for a calculation of all necessary parameters of this model. The macroscopic charge carrier mobility of the material was estimated by properly averaging the rate constants over the total simulation cell [136].

The electronic coupling matrix element responsible for positive charge carrier (hole) transport has been calculated for a variety of 4,4′-bis(9-carbazolyl)-biphenyl dimers corresponding to a disordered solid matrix. A procedure applying the computational scheme to a real system (organic molecules with a large molecular mass) is proposed. The consistency of the results with the existing model concept of hopping conductivity in organic semiconductors is discussed [137].

A procedure for calculating the microscopic ET kinetics in disordered solid matrices was formulated. The procedure was oriented toward applications for estimating the mobility of charge carriers in modern photovoltaic devices. The proposed model for the elementary act of CT deliberately excludes the effect that polarization modes of a medium had on its kinetics. The model was based on the evolution of a complex of local molecular modes polarized by an excessive molecular charge. The relaxation of this subsystem when interacting with acoustic phonon modes resulted in energy exchange between a local reaction center and the medium. Two alternative kinetic regimes were observed: a high-temperature regime corresponding to the jump dissipative reaction mechanism and a low-temperature regime corresponding to coherent ET. A comprehensive description of the regions of transition between these mechanisms was impossible using this model. Efficient algorithms for calculating the rate constant were developed for each mechanism [138].

An algorithm for a theoretical calculation of transfer reaction rates for light quantum particles (i.e., electron and H atom transfers) in nonpolar solid matrices was formulated and justified. The mechanism postulated involved a local mode (either an intra- or an intermolecular one) serving as a mediator which accomplishes the energy exchange between the reacting high-frequency quantum mode and the phonon modes belonging to the environment. This approach uses as a background the Fermi golden rule beyond the usually applied spin-boson approximation. The dynamical treatment rests on the 1D version of the standard quantum relaxation equation for the reduced density matrix, which describes the frequency fluctuation spectrum for the local mode under consideration. The temperature dependence of a reaction rate is controlled by the dimensionless parameter $\xi(0) = h\omega_0/k_B T$, where ω_0 is the frequency of the local mode and T is the temperature. The realization of the computational scheme is different for the high/intermediate ($\xi(0) <$ 1–3) and for low ($\xi(0) >> 1$) temperature ranges. For the first (quasi-classical) kinetic regime, the Redfield approximation to the solution of the relaxation equation proved to be sufficient and efficient in practical applications. The study of the essentially quantum-mechanical low-temperature kinetic regime in its asymptotic limit requires the implementation of the exact relaxation equation. The coherent mechanism providing a nonvanishing reaction rate has been revealed when $T \to 0$. An accurate computational methodology for the cross-over kinetic regime needs a further elaboration. The original model of the hopping mechanism for electronic conduction in photosensitive organic materials is considered, based on the above techniques. The ET in active centers of such systems proceeds via local intra- and intermolecular modes. The active modes, as a rule, operate beyond the kinetic regimes, which are usually postulated in the existing theories of the ET. Our alternative dynamic ET model for local modes immersed in the continuum harmonic medium is formulated for both classical and quantum regimes, and accounts explicitly for the mode/medium interaction. The kinetics of the energy exchange between the local ET subsystem and the surrounding environment essentially determine the total ET rate. The efficient computer code for rate computations is elaborated on. The computations are available for a wide range of system parameters, such as the temperature, external field, local mode frequency, and characteristics of mode/medium interaction. The relation of the present approach to the Marcus ET theory and to the quantum-statistical reaction rate theory underlying it is discussed and illustrated by the results of computations for practically important target systems [135].

Nitranilic acid (2,5-dihydroxy-3,6-dinitro-2,5-cyclohexadiene-1,4-dione) as a strong dibasic acid in acidic aqueous media creates the Zundel cation, H_5O_2+. The structural unit in a crystal comprises (H_5O_2)(2)(+)(2,5-dihydroxy-3,6-dinitro-1,4-benzoquinonate) (2-dihydrate) where the Zundel cation reveals no symmetry, being an ideal case for studying proton dynamics and its stability. The Zundel cation and PT dynamics are studied by variable-temperature X-ray diffraction, IR and solid-state NMR spectroscopy, and various quantum chemical methods, including periodic DFT calculations, ab initio MD simulation, and quantization of nuclear motion along three fully coupled internal coordinates. The Zundel cation features a short H-bond with the O···O distance of 2.433(2) Å with an asymmetric placement of hydrogen. The proton potential is of a single well type and, due to the nonsymmetric surroundings, of asymmetric shape. The formation of the Zundel cation is facilitated by the electronegative NO_2 groups. The employed spectroscopic techniques supported by calculations confirm the presence of a short H-bond with a complex proton dynamics [139].

Outer-sphere reorganization energies (λ) for intramolecular electron and hole transfer were studied in anion and cation radical forms of complex organic substrates (*p*-phenylphenyl-spacer-naphthyl) in polar (water, 1,2-dichloroethane, tetrahydrofuran) and quadrupolar (supercritical CO_2) solvents. The structure and charge distributions of solute molecules are obtained at the HF/6-31G(d,p) level. Standard LJ parameters for solutes and the nonpolarizable simple site-based models of solvents are used in MD simulations. Calculation of λ is done by means of the original procedure, which treats electrostatic polarization of a solvent in terms of a usual nonpolarizable MD scheme supplemented by scaling of reorganization energies at the final stage. This approach provides a physically relevant background for separating inertial and inertialess polarization responses by means of a single parameter $\varepsilon(\infty)$, optical dielectric permittivity of the solvent. Absolute λ values for hole transfer in 1,2-dichloroethane agree with results of previous computations in terms of the different technique (MD/FRCM [140]). Computed λ values for ET in tetrahydrofuran are larger than the experimental values by ~2.5 kcal/mol; for the case of hole transfer in 1,2-dichloroethane the discrepancy is of similar magnitude provided the experimental data are properly corrected. The MD approach gives nonzero λ values for the CT reaction in supercritical CO_2, being able to provide a uniform treatment of nonequilibrium solvation phenomena in both quadrupolar and polar solvents [125].

Outersphere reorganization energies (λ) for intramolecular ET and hole transfer were studied in anion- and cation-radical forms of complex organic substrates (biphenylyl-spacer-naphtyl) in polar solvents simulated by means of the nonpolarizable models of water and 1,2-dichloroethane. The earlier elaborated molecular/continuum approach (MD/FRCM [89]) was used; this method provides a physically relevant background for separating inertial and inertialess polarization responses within a nonpolarizable MD simulation (the SPC water model). QC calculations of solute charge distributions were performed with semiempirical (AM1) and ab initio (HF/6-31G(d,p)) approximations. Ab initio charges gave lower λ values and are preferable, probably, because of including the effect of the self-consistent reaction field (SCRF) polarization of the diabatic ET states. Standard LJ and charge parameters implemented in MD runs were not specially fitted for reproducing ET effects. The difference in values for a cation and an anion originating from the same parent structure was specially investigated. As shown earlier, this effect, nonlinear in its nature, proved to be extremely large when a model dipolar two-site system was studied. For the present ET structures representing real chemical substrates it has reduced to a plausible value of 6–8 kcal/mol. The study of the temperature dependence of A comprises a first MD simulation of this problem and its slope was found to be in accord with an experimental observation for an anionic species. Calculations of absolute λ values for the hole transfer in 1,2-dichloroethane are the first MD simulations of reorganization energies in experimentally studied reactions. Computed λ values are higher than the experimental data. This effect could be eliminated by properly tuning the solvent parameters [140].

A method of calculation of a free-energy surface (FES) of the PT reaction in a polar aprotic solvent was developed. This method was based on the two-state (valence bond) VB description of the solute combined with recent continuum medium models. Its essential new feature was an explicit QC treatment of VB wave functions, including internal electronic structure of a chemical subsystem. The FES includes a pair of intrasolute coordinates, *R*, the distance between hydrogen-bonded atoms and *s*, the proton coordinate, together with the collective medium polarization mode. Two hydrogen-bonded systems immersed in a polar solvent (Freon) were considered. The first one is the H(5)O(2)(+) ion, a model system which was used as a benchmark testifying the validity of these semiempirical calculations. The second system was the neutral (CN)(CH(3))N–H···N(CH(3))(3) complex in Freon. The PT for this system was studied experimentally. The dependencies of basic parameters controlling FES properties (the overlap integral, the coupling matrix element, and the reorganization energy E_r) on intrasolute coordinates *R* and *s* were evaluated and discussed. In particular, for the neutral complex, E_r, depended on *s* linearly, and its dependence on *R* was weak. The FES, for the neutral system, has two potential wells separated by the energy barrier of similar to 7 kcal/mol. Quantum mechanical averaging over the proton coordinate, *s*, reduces the barrier from 7.0 to 1.2 kcal/mol. The value of the nonadiabatic parameter on the averaged FES is equal to 0.13. This implies that the PT in the second system corresponds to an intermediate dynamic regime and that proton tunneling effects are hardly significant for this reaction [141].

Model studies were made of the nontransferred (reactant) and proton-transferred (product) species of the ground-state hydrogen-bonded clusters ROH···(NH3)(5), where ROH is an aromatic alcohol. The structure, relative energies and vibrational frequencies of this species were obtained at the HF/6-31G* level of theory. It was found that PT could proceed in the case of phenols with the electron-withdrawing substituents. A low-frequency vibration, which corresponds to the synchronous motion of the solvation shell, does exist in these clusters. The energy of the lowest π–π* transition is calculated for aromatic alcohols and their anions at the CASPT2 level with a VDZ basis set [142].

The quantum dynamics of a PT reaction in a polar solvent, treated as continuum, was considered taking as an example the PT process Flu(−) + HFlu → HFlu + Flu(−) in ether, where FluH means fluorene. Using the model 3D FES derived from QC SCRF calculations, the dynamical description was reduced to a two-level stochastic Liouville equation in the 2D subspace spanned by the solute vibrational mode (representing a relative motion of heavy atoms constituting the PT reaction center) and a solvent collective coordinate. The two quantum states involved in a reactive event were a pair of lowest proton levels obtained by means of quantum-mechanical averaging of the basic 3D FES. The new methodology of a direct evaluation of the coupling matrix element was elaborated. The rate calculation involved a treatment of extremely small (similar to 10^{-5}–10^{-10}) transmission factors for which two different approximate nonadiabatic approaches were tested. The whole variety of experimental data on reaction (a) involving both the absolute values of the rate constant (K-H) and the H/D isotope effect cannot be consistently described within the present two-level scheme [143].

The PT reaction R(−) + HR → RH + R(−) of benzyl-type compounds in a polar solvent was studied theoretically in terms of a proton adiabatic dynamical treatment of a mixed quantum-classical reacting system. The gas-phase potentials were obtained using the PM3 method and then approximated by appropriate (LEPS-type) quasi-analytical functions. Polar medium degrees of freedom were introduced, similar to Marcus electron transfer theory, in terms of a two-state electronic Hamiltonian. At this level, 3D FESs were obtained, including a pair of intrasolute coordinates, the C–H stretch and heavy-atom vibrational mode of the reaction center, together with the collective medium polarization mode. At the next stage, 2D electron-proton free-energy surfaces (EP FESs) corresponding to the adiabatic approximation with respect to C–H stretch were generated for the two lowest proton levels. Their main features are described. The reaction with R = benzyl proved to be proton-adiabatic. Its rate constant transmission factor calculated in terms of the Kramers–Grote–Hynes theory is significantly less than unity (similar to 0.4–0.6) because the reaction coordinate at the transition state of the ground state EP FES coincides with the medium mode. The reaction with R = fluorenyl does not obey the proton adiabaticity condition and needs a special kinetic treatment. A remarkable observation is that the double adiabatic electron-proton approximation was incapable of providing sufficiently high classical barriers (>10 kcal/mol) on ground-state 2D EP FESs for PT reactions in polar solvents [144].

The QC method AM1 was first used for investigating the processes of gas-phase dissociative electron transfer (DET) to the molecules of alkyl halides RX. The investigated alkyl halides included chlorine, bromine, and iodine derivatives, as well as polyhalogenated compounds. The AM1 method was found to yield realistic results that agree with experimental data. According to the postulate of the GM model (the gas-phase model with a predissociation minimum) suggested previously, a predissociation minimum for most of the radical anions [R-X]radical anion was detected on the potential energy curve. By making the AM1 calculation, the effective activation energies of ET ($E_{(ae)^*}$), the bond rupture in the radical anion (E_{ar}), and the values of intramolecular energy of reorganization λ_i were estimated. Certain regularities resulting from the GM-model were established. These are: the $E_{(ae)^*}$ decrease in the series X Cl > Br > I; the intibate change in the activation energies $E_{(ae)^*}$ and E_{ar}; and the linear relationship between force constants *f* and *f* of the R–X bond in the starting molecule and radical anions, with *f'* being invariably less than *f* [145].

The 2D potential energy surface (PES) of the quasi-symmetric OHO fragment was used for treatment of PT dynamics. The PES described semiquantitatively the main experimental regularities for a strong hydrogen bond. Strong coupling of modes (proton motion and O···O vibration) and dynamic asymmetry of the PES were taken into account. The respective wave equation was solved numerically with adiabatic separation of the fast (proton) and slow (O···O vibration) subsystems. Quantum jumps between vibrational levels of both subsystems under the random force action of the environment were assumed to accomplish PT from one well into another. The tunnel transitions between the slow subsystem levels, corresponding to the proton localization in different wells, were most important, their probabilities depending strongly on the O···O equilibrium separation. At large O···O distances (almost equal to 2.64 Å) the total tunnel transition probability from definite O···O vibrational levels (*m*) of one well into all possible levels of the other well is shown to increase with *m*. Such a promotion of the proton tunneling was observed by several authors at the laser selective vibrational excitation of the slow subsystem. For smaller O···O separation (about 2.52 Å) no vibrational assistance of the proton tunneling occurs. The microscopic mechanism of the process for these two cases was interpreted in terms of the respective matrix elements [146].

The solvent reorganization energies λ_0 of the electron self-exchange reaction between neutral molecules (nitrobenzene, its five *p*-substituted derivatives and benzonitrile) and the corresponding radical anions are discussed in terms of the Kirkwood approach combined with a semiempirical calculation of the reactant structure (AM1 and intermediate neglect of differential overlap [INDO] methods). The theoretical values of λ_0 are compared with the experimental reorganization energies λ calculated from the Marcus equation and the homogeneous rate constants k_{ex}. It is found that the values of λ_0 obtained using the INDO method are approximately equal to the experimental λ, whereas the AM1 method gives poor results for *p*-substituted nitrobenzenes. The linear correlation of log k_{ex} with a_{N2} where a_N is a nitrogen coupling constant, established for *p*-substituted nitrobenzenes is interpreted. A simple criterion for slow ET of an arbitrary flat polar arene is formulated in terms of the reactant structure and the model of the specifying group suggested earlier by some of the authors [147].

The most important problems arising in the theory of collective medium coordinates (CMCs) were analyzed. A CMC was defined as a highly correlated motion of a large ensemble of medium particles, which, as a single variable, underwent a significant change during a chemical process proceeding in this medium. The concept of CMCs was considered efficacious in the case of CT reactions in polar solvents. The following problems of CMC theory were considered: (i) quantum chemical computations of FESs explicitly treating a CMC as an essential coordinate, including location of minimum (reactants and products) and saddle (transition states) points on these surfaces; (ii) derivation of stochastic dynamical equations for CMCs and evaluation of their parameters, such as masses, friction coefficients, etc.; and (iii) derivation of the rate expression for stochastic dynamical systems, including CMCs. These problems were discussed using as examples electron and proton transfer and

S(N)2 reactions. A qualitative criterion was formulated for the case when an explicit consideration of a CMC was necessary and when it could be discarded [148].

The potential energy surface of the ground stare and lowest lying singlet of *o*-hydroxyacetophenone was calculated at the ab initio self-consistent field (SCF) and singly excited configuration interaction (CIS) levels, respectively. A split-valence basis set was used, to which polarization functions on atoms involved in the intramolecular H-bond were added. Whereas all these potentials contained one minimum corresponding to the normal form and a second well for the proton-transferred tautomer, only one minimum appeared in these potentials after correlation was added. The $S_0 \rightarrow S_1$ excitation induced the tautomerization reaction associated with intramolecular PT from the hydroxyl to the carbonyl oxygen. The excited-state PT might involve a nonradiative transition from the $S_1(n\ \pi^*)$ state to $S_1(\pi\ \pi^*)$. The vibrational spectra of the two tautomers exhibited strong differences. An anharmonic adiabatic treatment of the OH···O segment of the molecule led to an improved reproduction of experimentally observed frequencies of (O···O) and (OH) vibrational bands [149].

References

1. A. Bagaturyants and M. Alfimov, Atomistic simulation of hierarchical nanostructured materials for optical chemical sensing, in *Chemical Sensors: Simulation and Modeling*, Vol. 4, *Optical Sensors*, G. Korotcenkov (ed.), Momentum Press, New York, 2013, pp. 1–38.
2. F. Baldini, A. N. Chester, J. Homola, and S. Martellucci, *Optical Chemical Sensors*, Springer-Verlag, Dordrecht, the Netherlands, 2006.
3. C. McDonagh, C. S. Burke, and B. D. MacCraith, Optical chemical sensors, *Chem. Rev.*, 2008, **108**, 400–422.
4. W. R. Seitz, Chemical sensors based on fiber optics, *Anal. Chem.*, 1984, **56**. 16A–34A.
5. L. L. Blyler, Jr., R. A. Lieberman, L. G. Cohen, J. A. Ferrara, and J. B. MacChesney, Optical fiber chemical sensors utilizing dye-doped silicone polymer claddings, *Polym. Eng. Sci.*, 1989, **29**, 1215–1218.
6. O. S. Wolfbeis, Optical sensing based on analyte recognition by enzymes, carriers and molecular interactions, *Anal Chim. Acta*, 1991, **250**, 181–201.

7. D. Crowther and X. Liu, Covalent immobilisation of solvatochromic dyes, *J. Chem. Soc. Chem. Commun.*, 1995, **1995**, 2445.
8. O. S. Wolfbeis, Materials for fluorescence-based optical chemical sensors, *J. Mater. Chem.*, 2005, **15**, 2657–2669.
9. T. A. Dickinson, J. White, J. S. Kauer, and D. R. Walt, A chemical-detecting system based on a cross-reactive optical sensor array, *Nature*, 1996, **382**, 697–700.
10. M. F. Choi and P. Hawkins, The development of optical chemical sensors for the detection of volatile compounds from spoiled hams, *Sens. Actuators B*, 1997, **38–39**, 390–394.
11. R. Paolesse, D. Monti, F. Dini, and C. Di Natale, Fluorescence based sensor arrays, *Top. Curr. Chem.*, 2011, **300**, 139–174.
12. A. Khlebunov, D. Ionov, V. Aristarkhov, V. Sazhnikov, and M. Alfimov, A study of optical chemosensor materials, *Chem. Eng. Trans.*, 2008, **15**, 293–300.
13. M. V. Alfimov, A. A. Bagatur'yants, A. A. Safonov, A. V. Scherbinin, K. G. Vladimirova, S. A. Belousov, M. V. Bogdanova, I. A. Valuev, A. V. Deinega, Yu. E. Lozovik, and B. V. Potapkin, Multiscale computer design of photonic crystal based materials for optical chemosensors, *Nanotechnol. Russ.*, 2010, **5**, 250–258.
14. R. Phillips, Multiscale modeling in the mechanics of materials, *Curr. Opin. Solid State Mater. Sci.*, 1998, **3**, 526–532.
15. R. M. Nieminen, From atomistic simulation towards multiscale modelling of materials, *J. Phys.: Condens. Matter*, 2002, **14**, 2859–2876.
16. J. A. Elliott, Novel approaches to multiscale modelling in materials science, *Int. Mater. Rev.*, 2011, **56**, 207–225.
17. K. Ya. Burshtein, A. A. Bagatur'yants, and M. V. Alfimov, Shapes of absorption bands in solutions: organic chromophores, *J. Mol. Struct. THEOCHEM*, 1994, **314**, 311–320.
18. K. Ya. Burshtein, A. A. Bagatur'yants, and M. V. Alfimov, MO calculations on the absorption spectra of organic dimers. The interaction energy between dipole moments of electronic transitions in monomers and the shape of absorption bands, *Chem. Phys. Lett.*, 1995, **239**, 195–200.
19. K. Ya. Burshtein, A. A. Bagatur'yants, and M. V. Alfimov, Computer simulation of the shape of absorption bands in electronic spectra of J-aggregates, *Russ. Chem. Bull.*, 1997, **46**, 62–64.
20. A. A. Bagatur'yants, A. A. Korkin, K. P. Novoselov, L. L. Savchenko, and S. Ya. Umanskii, Integrated approach to atomistic simulation of film

deposition processes, in *Computational Materials Science*, C. R. A. Catlow and E. Kotomin (eds.), IOS Press, Amsterdam, 2003, 388–418.

21. A. A. Bagatur'yants, A. Kh. Minushev, K. P. Novoselov, A. A. Safonov, S. Ya. Umanskii, A. S. Vladimirov, and A. Korkin, Atomistic simulation of silicon nitride CVD from dichlorosilane and ammonia, in *Predictive Simulation of Semiconductor Processing Status and Challenges*, Springer Series in Materials Science, 72, J. Dabrowski and E. R. Weber (eds.), Springer Verlag, Berlin Heidelberg, 2004, 295–356.
22. A. A. Bagatur'yants, M. A. Deminskii, A. A. Knizhnik, B. V. Potapkin, and S. Ya. Umanskii, Integrated approach to dielectric film growth modeling: growth mechanisms and kinetics, in *Thin Films and Nanostructures: Physico-Chemical Phenomena in Thin Films and at Solid Surfaces*, L. I. Trakhtenberg, S. H. Lin, and O. J. Ilegbusi (eds.), Elsevier, Amsterdam, 2007, 468–522.
23. A. P. De Silva, H. Q. N. Gunaratne, T. Gunnlaugsson, A. J. M. Huxley, C. P. McCoy, J. P. Rademacher, and T. E. Rice, Signaling recognition events with fluorescent sensors and switches, *Chem. Rev.*, 1997, **97**, 1515–1566.
24. M. Schluter, J. R. Chelikowsky, S. G. Louie, M. L., and Cohen, Self-consistent pseudopotential calculations for Si (111) surfaces: unreconstructed (1x1) and reconstructed (2x1) model structures, *Phys. Rev. B*, 1975, **12**, 4200–4214.
25. F. Bechstedt, *Principles of Surface Physics*, Springer-Verlag, Berlin, 2003.
26. C. Lavor, A deterministic approach for global minimization of molecular potential energy functions, *Int. J. Quant. Chem.*, 2003, **95**, 336–343.
27. T. A. Halgren, Merck molecular force field: I. Basis, form, scope, parameterization, and performance of MMFF94, *J. Comput. Chem.*, 1996, **17**, 490.
28. F. V. Grigor'ev, A. N. Romanov, D. N. Laikov, S. N. Zhabin, A. Yu. Golovacheva, I. V. Oferkin, A. V. Sulimov, M. V. Bazilevskii, A. A. Bagatur'yants, V. B. Sulimov, and M. V. Alfimov, Molecular modeling methods for supramolecular complexes: a hierarchical approach, *Nanotechnol. Russ.*, 2010, **5**, 290–298.
29. D. Cramer, Møller–Plesset perturbation theory: from small molecule methods to methods for thousands of atoms, *WIREs Comput. Mol. Sci.*, 2011, **1**, 509–530.
30. R. J. Bartlett, Coupled-cluster approach to molecular structure and spectra: a step toward predictive quantum chemistry, *J. Phys. Chem.*, 1989, **93**, 1697–1708.

31. F. M. Ytreberg, R. H. Swendsen, and D. M. Zuckerman, Comparison of free energy methods for molecular systems, *J. Chem. Phys.*, 2006, **125**, 184114.

32. B. R. Brooks, I. D. Jane, and M. Karplus, Harmonic analysis of large systems: methodology, *J. Comp. Chem.*, 2004, **16**, 1522–1542.

33. V. G. Plotnikov, V. A. Sazhnikov, and M. V. Alfimov, Intermolecular interactions and spectral and luminescent properties of optical molecular sensors, *High Energy Chem.*, 2007, **41**, 299–311.

34. A. A. Khlebunov, D. S. Ionov, P. V. Komarov, V. M. Aristarkhov, V. A. Sazhnikov, A. N. Petrov, and M. V. Alfimov, An experimental system for investigating the characteristics of optical sensor materials, *Instrum. Exp. Tech.*, 2009, **52**, 132–136.

35. V. Sazhnikov and M. Alfimov, Rational design of optical chemosensor materials, *Chem. Eng. Trans.*, 2008, **15**, 267–274.

36. A. Calderone, R. Lazzaroni, and J. L. Bredas, Geometric and electronic structure of the butadiene/benzene complex as a model for the polyene/graphite interface, *Synth. Met.*, 1996, **82**, 225–229.

37. A. Calderone, R. Lazzaroni, and J. L. Bredas, Geometric and electronic structure of alkane/benzene, ethylbenzene/benzene, and alkane/ethylbenzene complexes: towards the characterization of polymer alloy composites, *Synth. Met.*, 1998, **95**, 1–15.

38. L. D. Perez, E. Florez, J. E. Mark, and B. L. Lopez, Estimation of the interaction energy between small molecules and a silica model as an approach for predicting the interaction order between elastomers and silica, *Polym. Int.*, 2009, **58**, 811–816.

39. A. A. Safonov, E. A. Rykova, A. A. Bagaturyants, V. A. Sazhnikov, and M. V. Alfimov, Atomistic simulations of materials for optical chemical sensors: DFT-D calculations of molecular interactions between gas-phase analyte molecules and simple substrate models, *J. Mol. Model.*, 2011, **17**, 1855–1862.

40. A. A. Safonov, A. A. Bagaturyants, V. A. Sazhnikov, and M. V. Alfimov, Density functional calculations of 9-diphenylaminoacridine fluorescent indicator and its interactions with analyte molecules: I. Structures of complexes in the ground electronic states and absorption spectra, *High Energy Chem.*, 2011, **45**, 229–236.

41. A. A. Safonov, A. A. Bagaturyants, V. A. Sazhnikov, and M. V. Alfimov, Density functional calculations of 9-diphenylaminoacridine fluorescent indicator and its interactions with analyte molecules: II. Structures of complexes in the excited electronic states and emission spectra, *High Energy Chem.*, 2011, **45**, 265–272.

42. F. Neese, with contributions from U. Becker, D. Ganiouchine, S. Koßmann, T. Petrenko, C. Riplinger, and F. Wennmohs, *ORCA, an ab initio, Density Functional and Semiempirical Program Package*, University of Bonn, Germany, 2010.

43. S. Grimme, Accurate description of van der Waals complexes by density functional theory including empirical corrections, *J. Comput. Chem.*, 2004, **25**, 1463–1476.

44. S. Grimme, Semiempirical GGA-type density functional constructed with a long-range dispersion correction, *J. Comput. Chem.*, 2006, **27**, 1787–1799.

45. S. Grimme, J. Antony, T. Schwabe, and C. Muck-Lichtenfeld, Density functional theory with dispersion corrections for supramolecular structures, aggregates, and complexes of (bio)organic molecules, *Org. Biomol. Chem.*, 2007, **5**, 741–758.

46. S. Grimme, Density functional theory with London dispersion corrections, *WIREs Comput. Mol. Sci.*, 2011, **1**, 211–228.

47. V. A. Sazhnikov, A. G. Strukov, M. G. Stunzhas, S. P. Efimov, O. M. Andreev, and M. V. Alfimov, Photoinduced formation of 9-substituted acridine from diphenylamine and tetrabromomethane, *Dokl. Akad. Nauk SSSR*, 1986, **288**, 172–176.

48. V. A. Sazhnikov, A. A. Khlebunov, and M. V. Alfimov, Solvatofluorochromic properties of 2,7-dimethyl-9-(ditolylamino)acridine, *High Energy Chem.*, 2007, **41**, 25–28.

49. D. Jacquemin, E. A. Perpete, G. E. Scuseria, I. Ciofini, and C. Adamo, TD-DFT performance for the visible absorption spectra of organic dyes: conventional versus long-range hybrids, *J. Chem. Theory Comput.*, 2008, **4**, 123–135.

50. A. A. Granovsky, Firefly version 8, http://www.classic.chem.msu.su/gran/firefly/index.html.

51. M. W. Schmidt, K. K. Baldridge, J. A. Boatz, S. T. Elbert, M. S. Gordon, J. H. Jensen, S. Koseki, N. Matsunaga, K. A. Nguyen, S. Su, T. L. Windus, M. Dupuis, and J. A. Montgomery, *J. Comput. Chem.*, 1993, **14**, 1347–1363.

52. R. Baer, E. Livshits, and U. Salzner, Tuned range-separated hybrids in density functional theory, *Annu. Rev. Phys. Chem.*, 2010, **61**, 85–109.

53. P. S. Rukin, A. A. Safonov, and A. A. Bagaturyants, Calculations of complexes of 9-diphenylaminoacridine fluorescent indicator with analyte molecules using density functional theory with dispersion correction, *Nanotechnol. Russ.*, 2011, **6**(5–6), 298–302.

54. S. Bonacchi, D. Genovese, R. Juris, M. Montalti, L. Prodi, E. Rampazzo, M. Sgarzi, and N. Zaccheroni, Luminescent chemosensors based on silica nanoparticles, *Top. Curr. Chem.*, 2011, **300**, 93–138.

55. J. Liu, X. Yang, X. He, K. Wang, Q. Wang, Q. Guo, H. Shi, J. Huang, and X. Huo, Fluorescent nanoparticles for chemical and biological sensing, *Sci. China-Chem.*, 2011, **54**, 1157–1176.

56. M.-M. Ramón, S. Félix, H. Mandy, B. Mustafa, and R. Knut, Nanoscopic optical sensors based on functional supramolecular hybrid materials, *Anal. Bioanal. Chem.*, 2011, **399**, 55–74.

57. R. F. Minibaev, N. A. Zhuravlev, A. A. Bagatur'yantz, and M. V. Alfimov, An investigation into adsorption of simple organic and inorganic molecules on a silica gel surface, *Russ. Phys. J.*, 2009, **52**, 1164–1169.

58. J. P. Perdew, K. Burke, and M. Ernzerhof, Generalized gradient approximation made simple, *Phys. Rev. Lett.*, 1996, **77**, 3865–3868.

59. D. Vanderbilt, Soft self-consistent pseudopotentials in a generalized eigenvalue formalism, *Phys. Rev. B*, 1990, **41**, 7892–7895.

60. P. Giannozzi, S. Baroni, N. Bonini, M. Calandra, R. Car, C. Cavazzoni, D. Ceresoli, G. L. Chiarotti, M. Cococcioni, I. Dabo, A. Dal Corso, S. Fabris, G. Fratesi, S. de Gironcoli, R. Gebauer, U. Gerstmann, C. Gougoussis, A. Kokalj, M. Lazzeri, L. Martin-Samos, N. Marzari, F. Mauri, R. Mazzarello, S. Paolini, A. Pasquarello, L. Paulatto, C. Sbraccia, S. Scandolo, G. Sclauzero, A. P. Seitsonen, A. Smogunov, P. Umari, and R. M. Wentzcovitch, Quantum ESPRESSO: a modular and open-source software project for quantum simulations of materials, *J. Phys.: Condens. Matter*, 2009, **21**, 395502.

61. P. Vashishta, R. K. Kalia, J. P. Rino, and I. Ebbsjo, Interaction potential for SiO_2: a molecular-dynamics study of structural correlations, *Phys. Rev. B*, 1990, **41**, 12197–12209.

62. V. Chashchikhin, E. Rykova, and A. Bagaturyants, Density functional theory modeling of the adsorption of small analyte and indicator dye 9-(diphenylamino)acridine molecules on the surface of amorphous silica nanoparticles, *Phys. Chem. Chem. Phys.*, 2011, **13**, 1440–1447.

63. P. K. Jal, M. Sudarshan, A. Saha, S. Patel, and B. K. Mishra, Synthesis and characterization of nanosilica prepared by precipitation method, *Colloids Surf. A*, 2004, **240**, 173–178.

64. F. Tielens, C. Gervais, J.-F. Lambert, F. Mauri, and D. Costa, Ab initio study of the hydroxylated surface of amorphous silica: a representative model, *Chem. Mater.*, 2008, **20**, 3336–3344.

65. P. Ugliengo, M. Sodupe, F. Musson, I. J. Bush, R. Orlando, and R. Dovesi, Realistic models of hydroxylated amorphous silica surfaces and MCM-41 mesoporous material simulated by large-scale periodic B3LYP calculations, *Adv. Mater.*, 2008, **20**, 4579–4583.

66. V. Chashchikhin, E. Rykova, and A. Bagatur'yants, Effect of analyte molecules on the electronic absorption and fluorescence spectra of a receptor center based on the 9-(diphenylamino)acridine dye adsorbed on silica clusters, *Nanotechnol. Russ.*, 2011, **6**, 579–586.

67. V. A. Tikhomirov, A. V. Odinokov, A. A. Bagatur'yants, and M. V. Alfimov, Modeling the surface of polystyrene and the adsorption of dye molecules on this surface, *Theor. Experim. Chem.*, 2011, **46**, 342–349.

68. A. K. Dutta, K. Kamada, and K. Ohta, Spectroscopic studies of Nile red in organic solvents and polymers, *J. Photochem. Photobiol. A*, 1996, **93**, 57–64.

69. T. Behnke, C. Wurth, K. Hoffmann, M. Hubner, U. Panne, and U. Resch-Genger, Encapsulation of hydrophobic dyes in polystyrene micro and nanoparticles via swelling procedures, *J. Fluoresc.*, 2011, **21**, 937–944.

70. B. Hess, C. Kutzner, D. van der Spoel, and E. Lindahl, GROMACS 4: algorithms for highly efficient, load-balanced, and scalable molecular simulation, *J. Chem. Theory Comput.*, 2008, **4**, 435–447.

71. D. van der Spoel, E. Lindahl, B. Hess, A. R. van Buuren, E. Apol, P. J. Meulenhoff, D. P. Tieleman, A. L. T. M. Sijbers, K. A. Feenstra, R. van Drunen, and H. J. C. Berendsen, Gromacs User Manual version 4.5.4, 2010, www.gromacs.org.

72. W. L. Jorgensen, D. S. Maxwell, and J. Tirado-Rives, Development and testing of the OPLS all-atom force field on conformational energetics and properties of organic liquids, *J. Am. Chem. Soc.*, 1996, **118**, 11225–11236.

73. P. Doruker and W. L. Mattice, Simulation of polyethylene thin films on a high coordination lattice, *Macromolecules*, 1998, **31**, 1418–1426.

74. P. O. Tuck, R. C. Mawhinney, and M. Rappon, An ab initio and TD-DFT study of solvent effect contributions to the electronic spectrum of Nile red, *Phys. Chem. Chem. Phys.*, 2009, **11**, 4471–4480.

75. F. M. Richards, Areas, volumes, packing, and protein structure, *Annu. Rev. Biophys. Bioeng.*, 1977, **6**, 151–176.

76. A. Varshney, F. P. Brooks, and W. V. Wright, Computing smooth molecular-surfaces, *IEEE Comput. Graph. Appl.*, 1994, **14**, 19–25.

77. A. Freidzon, V. Tikhomirov, A. Odinokov, and A. Bagaturyants, Multiscale approach to the structure and spectra of Nile red adsorbed

on polystyrene nanoparticles, *IOP Conf. Ser.: Mater. Sci. Eng.*, 2012, **38**, 012036.

78. J. H. Jensen, P. N. Day, M. S. Gordon, H. Basch, D. Cohen, D. R. Garmer, M. Krauss, and W. J. Stevens, Effective fragment method for modeling intermolecular hydrogen bonding effects on quantum mechanical calculations, in *Modeling the Hydrogen Bond*, D. A. Smith (ed.), ACS Symposium Series, Washington DC, 1994, **569**, 139–151.

79. M. S. Gordon, D. G. Fedorov, S. R. Pruitt, and L. V. Slipchenko, Fragmentation methods: a route to accurate calculations on large systems, *Chem. Rev.*, 2012, **112**, 632–672.

80. P. V. Yurenev, M. K. Kretov, A. V. Scherbinin, and N. F. Stepanov, Environmental broadening of the CTTS bands: the hexaammineruthenium(II) complex in aqueous solution, *J. Phys. Chem. A*, 2010, **114**, 12804–12812.

81. S. I. Pekar, On the effect of lattice distortion by electrons on the optical and electrical properties of crystals, *Usp. Fiz. Nauk*, 1953, **50**, 197–252.

82. V. Chashchikhin, E. Rykova, A. Scherbinin, and A. Bagaturyants, DFT modeling of band shifts and widths in the absorption spectrum of a 9-(diphenylamino)acridine/silica receptor center upon its interaction with gas-phase NH_3, C_2H_5OH, and $(CH_3)_2CO$ molecules, *Int. J. Quantum Chem.*, 2012, **112**, 3110–3118.

83. M. V. Vener, How reliable is the Lippincott-Schroeder potential for the OH...N hydrogen-bonded fragment in the gas-phase, *Chem. Phys. Lett.*, 1995, **244**, 89–92.

84. V. P. Sakun, M. V. Vener, and N. D. Sokolov, Proton tunneling assisted by the intermolecular vibration excitation. Temperature dependence of the proton spinlattice relaxation time in benzoic acid powder, *J. Chem. Phys.*, 1996, **105**, 379–387.

85. M. V. Vener and N. D. Sokolov, On the adiabatic separation of the vibrational variables of a hydrogen-bonded AHA fragment with a symmetric double-well potential, *Chem. Phys. Lett.*, 1997, **264**, 429–434.

86. M. V. Basilevsky, D. F. Parsons, and M. V. Vener, An advanced dielectric continuum approach for treating solvation effects: time correlation functions. I. Local treatment, *J. Chem. Phys.*, 1998, **108**(3), 1103–1114.

87. D. F. Parsons, M. V. Vener, and M. V. Basilevsky, Advanced continuum approaches for treating time correlation functions. The role of solute shape and solvent structure, *J. Phys. Chem. A*, 1999, **103**, 1171–1178.

88. M. V. Vener, I. V. Leontyev, Yu. A. Dyakov, M. V. Basilevsky, and M. D. Newton, Application of the linearized MD approach for computing equilibrium solvation free energies of charged and dipolar solutes in polar solvents, *J. Phys. Chem. B*, 2002, **106**, 13078–13088.

89. I. V. Leontyev, M. V. Vener, I. V. Rostov, M. V. Basilevsky, and M. D. Newton, Continuum level treatment of electronic polarization in the framework of molecular simulations of solvation effects, *J. Chem. Phys.*, 2003, **119**, 8024–8031.

90. M. V. Vener, I. V. Leontyev, and M. V. Basilevsky, Computations of solvation free energies for polyatomic ions in water in terms of a combined molecular-continuum approach, *J. Chem. Phys.*, 2003, **119**, 8038.

91. M. V. Vener, A. N. Egorova, D. P. Fomin, and V. G. Tsirelson, QTAIM study of the closed-shell interactions in peptide secondary structures: a cluster treatment of oligo- and polyalanines, *Chem. Phys. Lett.*, 2007, **440**, 279–285.

92. M. V. Vener, A. V. Manaev, D. Hadži, and V. G. Tsirelson, DFT study of proton dynamics in the potassium hydrogen maleate crystal: the infrared versus the inelastic neutron scattering spectra, *Z. Phys. Chem.*, 2008, **222**, 1349–1358.

93. M. V. Vener, X. Rozanskay, and J. Sauer, Protonation of water clusters in the cavities of acidic zeolites: $(H_2O)n$•H-chabazite, n = 1–4, *Phys. Chem. Chem. Phys.*, 2009, **11**, 1702–1712.

94. M. V. Vener, A. N. Egorova, D. P. Fomin, and V. G. Tsirelson, Hierarchy of the non-covalent interactions in the alanine-based secondary structures. DFT study of the frequency shifts and electron-density features, *J. Phys. Org. Chem.*, 2009, **22**, 177–185.

95. M. V. Vener, A. N. Egorova, D. P. Fomin, and V. G. Tsirel'son, A quantum-topological analysis of noncovalent interactions in secondary polyalanine structures, *Russ. J. Phys. Chem. B*, 2009, **3**(4), 541–547.

96. M. V. Vener, A. N. Egorova, D. P. Fomin, and V. G. Tsirelson, DFT study of H-bonds in the peptide secondary structures: the backbone–side-chain and polar side-chains interactions, *J. Mol. Struct.*, 2010, **972**, 11–15.

97. E. V. Bartashevich, D. K. Nikulov, M. V. Vener, and V. G. Tsirelson, QTAIMC study of the X–H/H...O bond order indices (X = O, N, C) in molecular systems, *Comput. Theor. Chem.*, 2011, **973**, 33–39.

98. S. A. Katsyuba, M. V. Vener, E. E. Zvereva, Z. Fei, R. Scopelliti, G. Laurenczy, N. Yan, E. Paunescu, and P. J. Dyson, How strong is hydrogen bonding

in ionic liquids? Combined X-ray crystallographic, infrared/Raman spectroscopic, and density functional theory study, *J. Phys. Chem. B*, 2013, **117**, 9094–9105.

99. M. Basilevsky, A. Odinokov, E. Nikitina, F. Grigoriev, N. Petrov, and M. Alfimov, Advanced dielectric continuum model of preferential solvation, *J. Chem. Phys.*, 2009, **130**, 024505.

100. M. Basilevsky, A. Odinokov, E. Nikitina, F. Grigoriev, N. Petrov, and M. Alfimov, Preferential solvation of spherical ions in binary DMSO/benzene mixtures, *J. Chem. Phys.*, 2009, **130**, 024504.

101. A. V. Odinokov, I. V. Leontyev, M. V. Basilevsky, and N. Ch. Petrov, Potential of mean force for ion pairs in non-aqueous solvents. Comparison of polarizable and non-polarizable MD simulations, *Mol. Phys.: Int. J. Interface Chem. Phys.*, 2011, **109**(2), 217–227.

102. A. V. Odinokov, S. V. Titov, V. A. Tikhomirov, M. V. Basilevsky, and M. V. Alfimov, Inclusion complexes of β-cyclodextrine with organic ligands: molecular dynamics simulation of the thermodynamic stability in gas phase and in water solution, *Mol. Simul.*, 2013, **39**(6), 442–452.

103. M. Deminsky, A. Eletskii, A. Kniznik, A. Odinokov, V. Pentkovskii, and B. Potapkin, Molecular dynamic simulation of transmembrane pore growth, *J. Membr. Biol.*, 2013, **246**, 821–831.

104. M. V. Vener, I. G. Shenderovich, and A. A. Rykounov, A qualitative study of the effect of a counterion and polar environment on the structure and spectroscopic signatures of a hydrated hydroxyl anion, *Theor. Chem. Acc.*, 2013, **132**, 1361.

105. J. E., Donald, D. W. Kulp, and W. F. DeGrado, Salt bridges: geometrically specific, designable interactions, *Proteins*, 2011, **79**, 898–915.

106. M. V. Vener, A. V. Odinokov, C. Wehmeyer, and D. Sebastiani, The structure and IR signatures of the arginine-glutamate salt bridge. Insights from the classical MD simulations, *J. Chem. Phys.*, 2015, **142**(21), 215106.

107. A. Odinokov, A. Freidzon, and A. Bagaturyants, Molecular dynamics simulation of the glass transition in 4,4′-N,N′-dicarbazolylbiphenyl, *Chem. Phys. Lett.*, 2015, **633**, 41–46.

108. A. V. Odinokov and A. A. Bagaturyants, Specific interactions of neutral side chains of an adsorbed protein with the surface of alpha-quartz and silica gel, *J. Phys. Chem. B*, 2015, **119**, 8679–8684.

109. A. Odinokov and D. Ostroumov, Structural degradation and swelling of lipid bilayer under the action of benzene, *J. Phys. Chem. B*, 2015, **119**, 15006–15013.

110. M. V. Vener, Model study of proton transfer in a H-bonded cluster with an A–H···B reaction complex. Introduction of an effective coordinate for the solvation shell, *Chem. Phys.*, 1998, **233**, 77–83.

111. A. P. Voronin, G. L. Perlovich, and M. V. Vener, Effects of the crystal structure and thermodynamic stability on solubility of bioactive compounds: DFT study of isoniazid cocrystals, *Comp. Theor. Chem.*, 2015, **1092**, 1–11.

112. A. N. Manin, A. P. Voronin, A. V. Shishicina, M. V. Vener, A. V. Churakov, and G. L. Perlovich, Influence of secondary interactions on the structure, sublimation thermodynamics, and solubility of salicylate:4-hydroxybenzamide cocrystals. combined experimental and theoretical study, *J. Phys. Chem. B*, 2015, **119**, 10466–10477.

113. M. V. Vener, O. E. Levina, O. A. Koloskov, A. A. Rykounov, A. P. Voronin, and V. G. Tsirelson, Evaluation of the lattice energy of the two-component molecular crystals using solid-state DFT, *Cryst. Growth Des.*, 2014, **14**, 4997–5003.

114. M. V. Vener, A. V. Shishkina, A. A. Rykounov, and V. G. Tsirelson, Cl...Cl interactions in molecular crystals: insights from the theoretical charge density analysis, *J. Phys. Chem. A*, 2013, **117**(35), 8459–8467.

115. A. V. Shishkina, V. V. Zhurov, A. I. Stash, M. V. Vener, A. A. Pinkerton, and V. G. Tsirelson, Noncovalent interactions in crystalline picolinic acid N-oxide: insights from experimental and theoretical charge density analysis, *Cryst. Growth Des.*, 2013, **13**(2), 816–828.

116. M. V. Vener, A. N. Egorova, A. V. Churakov, and V. G. Tsirelson, Intermolecular hydrogen bond energies in crystals evaluated using electron density properties: DFT computations with periodic boundary conditions, *J. Comp. Chem.*, 2012, **33**(29), 2303–2309.

117. M. V. Vener, A. G. Medvedev, A. V. Churakov, P. V. Prikhodchenko, T. A. Tripol'skaya, and O. Lev, H-bond network in amino acid cocrystals with H_2O or H_2O_2. The DFT study of serine-H_2O and serine-H_2O_2, *J. Phys. Chem. A*, 2011, **115**, 13657–13663.

118. A. V. Churakov, P. V. Prikhodchenko, O. Lev, A. G. Medvedev, T. A. Tripol'skaya, and M. V. Vener, A model proton-transfer system in the condensed phase: NH4 +OOH−, a crystal with short intermolecular H-bonds, *J. Chem. Phys.*, 2010, **133**, 164506.

119. M. V. Vener, A. V. Manaev, and V. G. Tsirelson, Proton dynamics in strong (short) intramolecular H-bond. DFT study of the KH maleate crystal, *J. Phys. Chem. A*, 2008, **112**(51), 13628–13632.

120. M. V. Vener, A. V. Manaev, A. N. Egorova, and V. G. Tsirel'son, A quantum-topological analysis of short (strong) H bonds in three-dimensional periodic crystals, *Russ. J. Phys. Chem. B*, 2008, **2**(4), 512–524.

121. A. N. Egorova, M. V. Vener, I. S. Zhukov, G. D. Kozak, and V. G. Tsirelson, Molecular structure, barriers to internal rotation, and the enthalpy of formation of cumene hydroperoxide PhCMe2OOH: a DFT study, *Russ. Chem. Bull.*, 2008, **57**(6), 1179–1186.

122. M. V. Vener, A. V. Manaev, A. N. Egorova, and V. G. Tsirel'son, QTAIM study of strong H-bonds with the O-H...A fragment (A = O, N) in three-dimensional periodical crystals, *J. Phys. Chem. A*, 2007, **111**(6), 1155–1162.

123. M. V. Vener and J. Sauer, Vibrational spectra of the methanol tetramer in the OH stretch region. Two cyclic isomers and concerted proton tunneling, *J. Chem. Phys.*, 2001, **114**(6), 2623–2628.

124. G. V. Zakharova, R. R. Konstantinov, A. V. Odinokov, A. K. Chibisov, M. V. Alfimov, I. E. Kasheverov, Yu. N. Utkin, M. N. Zhmak, and V. I. Tsetlin, Effect of a peptide modeling the nicotinic receptor binding site on the spectral and luminescent properties of dye complexes with cucurbit[8]uril, *High Energy Chem.*, 2016, **50**(2), 121–126.

125. (a) M. V. Vener, A. V. Tovmash, I. V. Rostov, and M. V. Basilevsky, Molecular simulations of outersphere reorganization energies in polar and quadrupolar solvents. The case of intramolecular electron and hole transfer, *J. Phys. Chem. B*, 2006, **110**, 14950–14955; (b) F. Ingrosso et al., Solvation dynamics in acetonitrile: a study incorporating solute electronic response and nuclear relaxation,. *J. Phys. Chem. B*, 2005, **109**(8), 3553–3564.

126. E. A. Nikitina, A. V. Odinokov, F. V. Grigoriev, M. V. Basilevsky, A. A. Khlebunov, V. A. Sazhnikov, and M. V. Alfimov, Molecular simulation of solvent-induced Stokes shift in absorption/emission spectra of organic chromophores, *J. Phys. Chem. B*, 2007, **111**, 3953–3959.

127. E. O. Levina, B. V. Lokshin, B. D. Mai, and M. V. Vener, Spectral features of guanidinium-carboxylate salt bridges. The combined ATR-IR and theoretical studies of aqueous solution of guanidinium acetate, *Chem. Phys. Lett.*, 2016, **659**, 117–120.

128. V. D. Maiorov, I. S. Kislina, A. A. Rykounov, and M. V. Vener, The structure and vibrational features of proton disolvates in water–ethanol solutions of HCl: the combined spectroscopic and theoretical study, *J. Phys. Org. Chem.*, 2014, **27**, 135–141.

129. M. V. Vener, S. Kong, A. A. Levina, and I. G. Shenderovich, Spectroscopic signatures of [H9O4]+ and [H13O6]+ ions in a polar aprotic

environment revealed under DFT-PCM approximation, *Acta Chim. Slov.*, 2011, **58**(3), 402–410.

130. M. V. Vener, A. N. Egorova, and V. G. Tsirelson, Hydrogen bonds and O…O interactions in proton-ordered ices. DFT computations with periodic boundary conditions, *Chem. Phys. Lett.*, 2010, **500**, 272–276.

131. S. S. Kong, I. G. Shenderovich, and M. V. Vener, Density functional study of the proton transfer effect on vibrations of strong (short) intermolecular O-H…N/O···H-N+ hydrogen bonds in aprotic solvents, *J. Phys. Chem. A*, 2010, **114**, 2393–2399.

132. M. V. Vener, A. V. Manaev, D. Hadzi, and V. G. Tsirelson, DFT study of proton dynamics in the potassium hydrogen maleate crystal: the infrared versus the inelastic neutron scattering spectra, *Zeitschrift Phys. Chem.*, 2008, **222**, 1349–1358.

133. M. V. Vener, O. Kuhn, and J. M. Bowman, Vibrational spectrum of the formic acid dimer in the OH stretch region. A model 3D study, *Chem. Phys. Lett.*, 2001, **5–6**, 562–570.

134. M. V. Vener and J. Sauer, Quantum anharmonic frequencies of the O…H…O fragment of the H5O2+ ion: a model three-dimensional study, *Chem. Phys. Lett.*, 1999, **312**, 591–597.

135. M. V. Basilevsky, A. V. Odinokov, S. V. Titov, and E. A. Mitina, Golden rule kinetics of transfer reactions in condensed phase: the microscopic model of electron transfer reactions in disordered solid matrices, *J. Chem. Phys.*, 2013, **139**(23), 234102.

136. M. V. Basilevsky, A. V. Odinokov, and K. G. Komarova, Charge-transfer mobility parameters in photoelectronic devices: the advanced Miller–Abrahams computation, *J. Phys. Chem. B*, 2015, **119**(24), 7430–7438.

137. A. V. Odinokov and A. A. Bagaturyants, Calculation of electronic coupling matrix elements in 4,4′-bis(9-carbazolyl)-biphenyl amorphous material, *High Energy Chem.*, 2015, **49**(3), 173–176.

138. M. V. Basilevsky and A. V. Odinokov, Microscopic jump mechanism of electron conductivity in photosensitive materials, *Russ. J. Phys. Chem. A*, 2014, **88**(11), 1817–1825.

139. K. Molcanov, J. Stare, M. V. Vener, B. Kojic-Prodic, G. Mali, J. Grdadolnikbe, and V. Mohacek-Grosev, Nitranilic acid hexahydrate, a novel benchmark system of the Zundel cation in an intrinsically asymmetric environment: spectroscopic features and hydrogen bond dynamics characterised by experimental and theoretical methods, *Phys. Chem. Chem. Phys.*, 2014, **16**, 998.

140. I. V. Leontyev, A. V. Tovmash, M. V. Vener, I. V. Rostov, and M. V. Basilevsky, Molecular simulations of outersphere reorganization energies for intramolecular electron and hole transfer in polar solvents, *Chem. Phys.*, 2005, **319**, 4–15.

141. M. V. Vener, I. V. Rostov, A. V. Soudackov, and M. V. Basilevsky, Semiempirical modeling free energy surfaces for proton transfer in polar aprotic solvents, *Chem. Phys.*, 2000, **254**, 249–265.

142. M. V. Vener and S. Iwata, Model study of H-bonded ROH...(NH3)5 clusters: a search for possible ground-state proton transfer species, *Chem. Phys. Lett.*, 1998, **292**, 87–91.

143. M. V. Basilevsky, M. V. Vener, G. V. Davidovich, and A. V. Soudackov, Dynamics of proton transfer reactions in polar solvent in the non-adiabatic two-state approximation: test calculations for carbon-carbon reaction centre, *Chem. Phys.*, 1996, **208**, 267–282.

144. M. V. Basilevsky, A. V. Soudackov, and M. V. Vener, Electron-proton free-energy surfaces for proton-transfer reaction in polar-solvents: test calculations for carbon-carbon reaction centers, *Chem. Phys.*, 1995, **200**, 87–106.

145. M. V. Vener and V. G. Mairanovskii, Dissociative electron-transfer–gas-phase model with predissociation minimum (gm-model): consideration by AM1 method, *Russ. J. Electrochem.*, 1993, **29**(10), 1084–1089.

146. N. D. Sokolov and M. V. Vener, Proton tunneling assisted by the intermolecular vibration excitation in solid state, *Chem. Phys.*, 1992, **168**, 29–40.

147. M. V. Vener, A. V. Cheprakov, N. T. Ioffe, and V. G. Mairanovskii, Decisive-fragment model and theoretical calculations of solvent reorganization energies in symmetric electron exchange reactions in para-substituted nitrobenzenes, *Russ. J. Electrochem.*, 1992, **28**(9), 1079–1084.

148. M. V. Basilevsky and M. V. Vener, Collective medium coordinates and their application in the theory of chemical reactions, *J. Mol. Struct. THEOCHEM*, 1997, **398–399**, 81–92.

149. M. V. Vener and S. Scheiner, Hydrogen-bonding and proton-transfer in the ground and lowest excited singlet-states of o-hydroxyacetophenone, *J. Phys. Chem.*, 1995, **99**, 642–649.

Index

For Product Safety Concerns and Information please contact our EU
representative GPSR@taylorandfrancis.com
Taylor & Francis Verlag GmbH, Kaufingerstraße 24, 80331 München, Germany

www.ingramcontent.com/pod-product-compliance
Ingram Content Group UK Ltd.
Pitfield, Milton Keynes, MK11 3LW, UK
UKHW022159190926
475702UK00003B/101

* 9 7 8 9 8 1 4 7 7 4 4 0 6 *